兵工遗产
南京晨光1865创意产业园环境设计

CIID "室内设计 6+1" 2015（第三届）校企联合毕业设计

THE ORDANCE FACTORY RELIC
The Environmental Design of Nanjing Chenguang 1865 Creativity Industry Park

中国建筑学会室内设计分会
北京建筑大学   编

内 容 提 要

目前，国际社会对于工业遗产保护逐渐形成良好氛围，越来越多的国家开始重视保护工业遗产，在制定保护规划的基础上，通过合理利用使工业遗产的重要性得以最大限度的保存和再现，增强公众对工业遗产的认识。为探索"卓越工程师教育培养计划"目标下的室内建筑师教育培养之路，探讨历史文化名城近代工业遗产保护与利用问题，由中国建筑学会室内设计分会主办了CIID"室内设计6+1"2015（第三届）校企联合毕业设计活动，同济大学、华南理工大学、哈尔滨工业大学、西安建筑科技大学、北京建筑大学、南京艺术学院、浙江工业大学等作为参加高校，南京观筑历史建筑文化研究院作为"兵工遗产——南京晨光1865创意产业园环境设计"联合毕业设计命题和总指导企业。来自建筑学、工业设计、环境设计、艺术与科技等4个专业2015届毕业班师生们，就南京晨光1865创意产业园环境设计的工业遗产与室内设计、工业遗产与展示设计、工业遗产与景观设计等3个设计专题，联合开展了综合性实践教学活动。

据此，中国建筑学会室内设计分会和北京建筑大学联合编辑了《兵工遗产·南京晨光1865创意产业园环境设计——CIID"室内设计6+1"2015（第三届）校企联合毕业设计》，采用中英文对照方式记载了活动的命题研讨、开题仪式与现场踏勘、中期检查、答辩评审与表彰奖励、编辑出版、专题展览、对外交流等的过程印记。本书以6+1个"活动脚步"板块作为编辑体例，图文并茂，内容详实。本书可供建筑学、历史建筑保护工程、城乡规划、风景园林、工业设计、环境设计、产品设计、视觉传达设计、公共艺术、艺术与科技等专业人员及设置相关专业的院校师生参考借鉴。

出版资助：北京建筑大学北京市财政专项
学科建设——硕士学位授权学科点——设计学（含工业设计工程）
专业建设——工业设计、环境设计

图书在版编目（CIP）数据

兵工遗产·南京晨光1865创意产业园环境设计：CIID"室内设计6+1"2015（第三届）校企联合毕业设计 / 中国建筑学会室内设计分会，北京建筑大学编. — 北京：中国水利水电出版社，2015.10
ISBN 978-7-5170-3717-0

Ⅰ.①兵… Ⅱ.①中… ②北… Ⅲ.①室内装饰设计－环境设计－作品集－中国－现代 Ⅳ.①TU238

中国版本图书馆CIP数据核字(2015)第234478号

书　　名	兵工遗产·南京晨光1865创意产业园环境设计 ——CIID"室内设计6+1"2015（第三届）校企联合毕业设计
作　　者	中国建筑学会室内设计分会　北京建筑大学　编
出版发行	中国水利水电出版社 （北京市海淀区玉渊潭南路1号D座　100038） 网址：www.waterpub.com.cn E-mail：sales@waterpub.com.cn 电话：（010）68367658（发行部）
经　　售	北京科水图书销售中心（零售） 电话：（010）88383994、63202643、68545874 全国各地新华书店和相关出版物销售网点
排　　版	中国建筑学会室内设计分会
印　　刷	北京博图彩色印刷有限公司
规　　格	210mm×285mm　16开本　17印张　592千字
版　　次	2015年10月第1版　2015年10月第1次印刷
印　　数	0001—1000册
定　　价	120.00元

凡购买我社图书，如有缺页、倒页、脱页的，本社发行部负责调换

版权所有·侵权必究

编委会
Editorial Committee

Director   主任
Zou Huying, Li Aiqun, Wang Su   邹瑚莹 李爱群 汪 苏

Deputy Director   副主任
Ye Hong, Su Dan, Li Zhenning   叶 红 苏 丹 李振宁
Sun Yimin, Mei Hongyuan, Lin Baogang   孙一民 梅洪元 蔺宝钢
Liu Linan, Li Yiwen, Chang Hong   刘临安 李亦文 常 虹

Committee member   委员
Chen Weixin, Zou Chao, Zuo Yan   陈卫新 邹 超 左 琰
Jiang Wenyi, Chen Jianhua, Zhou Lijun   姜文艺 陈建华 周立军
Ma Hui, Liu Jie, Liu Xiaojun   马 辉 刘 杰 刘晓军
He Fangyao, Yang Lin, Zhu Ningke   何方瑶 杨 琳 朱宁克
Zhu Fei, Lv Qinzhi, Huang Yan   朱 飞 吕勤智 黄 焱

Chief Editor   主编
Chen Jingyong   陈静勇

Associate editor   副主编
Yang Lin, Zhu Ningke   杨 琳 朱宁克

Editorial Committee   编委
Pan Xiaowei   潘晓微

Graphic Design   装帧设计
Wang Zhaowei, Du Lijuan, Zhang Jinghui   王罂伟 杜丽娟 张婧翬

Translator   翻译
Wang Ziyue, Yu Yueyue   王子岳 于玥玥

Preface

Since 2013, the CIID "Interior Design 6+1" School and Enterprise Joint Graduation Design, which was hosted by CIID and created by universities and well-known designing enterprises together, has been held for three years successfully. For the three years, we take the topics of "Postgame Business Opportunities------ the postgame rebuilding of the National Stadium interior design" "The Volume of the New Environment of City Track, the environmental design of Shanghai Subway rebuilding" and "The Heritage of War Industry, the environmental design of Nanjing Chenguang 1865 Create Industry Park" as the theme of the 6+1 graduation design. Although the design topics are hypotheses, they are current practical problems that these well-known enterprises need to solve. The graduation designs not only face to the society, the market, but also make the students acquire practical knowledge before graduation. It will lay a good foundation for them to find their feet in future work as soon as possible and fit the high speed development of the society.

Another distinct feature of "the interior design 6+1" school and enterprise joint graduation design is creates an educational platform formed by universities and well-known designing enterprises together. The platform is participated by 6 universities. It is a platform for the teachers and students of 6 universities to communicate learn and improve, for the development of interior design education.

The CIID "Interior Design 6+1" School and Enterprise Joint Graduation Design achieved good results during the last three years. It helps to train qualified people of innovation and pioneering and push ahead the reform of university education. As the only academic community of China interior design, CIID promotes the development of our country's interior design industry. We value young students' development. Not for fame and gain, keeping steady and persistent hard working is our duty. In the future, we will sum up experiences and continue trying to making this activity better and promoting the reform and development of interior design education.

序

邹瑚莹
中国建筑学会室内设计分会理事长
清华大学教授

Zou Huying
President of CIID
Professor of Tsinghua University

自 2013 年以来，由中国建筑学会室内设计分会主办、高校与知名设计企业共同创建的 CIID"室内设计 6+1"校企联合毕业设计活动，至今年已经成功举办 3 年了。3 年来，分别以"赛后商机——国家体育场赛后改造室内设计"、"城轨新境——上海地铁改造环境设计"、"兵工遗产——南京晨光 1865 创意产业园环境设计"为题，开展 6+1 毕业设计。设计的题目虽然是真题假作，但题目的选定来自知名设计企业当前需要解决的实际课题，既面向社会，面向市场，又让学生在毕业之前能理论联系实际地掌握知识，为今后尽快适应工作打下基础，以适应当前社会的高速发展。

"室内设计 6+1"校企联合毕业设计的又一特色是打造了一个高校与知名设计企业共同创建的教育平台，一个有 6 所高校共同参与的平台。这是一个 6 校师生交流的平台，一个 6 校师生相互学习的平台，一个 6 校师生共同提高的平台，一个推动室内设计教育发展的平台。

"室内设计 6+1"校企联合毕业设计 3 年来取得了良好的成果。这有助于创新、创业型人才的培养，有助于学校教育改革的推进。CIID 作为中国室内设计界唯一的学术团体，推动我国室内设计事业的发展，重视年青学生的成长，不为虚名浮华，踏踏实实坚持不懈地努力工作是我们应尽的责任。今后，我们还要总结经验，不断摸索，把这项活动更好地开办下去，以促进室内设计教育的改革与提高。

Forword
Discussion on the Protection and Utilization of Industrial Design Innovation on Cultivation of Interior Design Talent Education

Guided by the strategy of building an innovative country and strengthening China by talents, based on the scientific and professional new environment, and facing the new target of cultivating interior architects, in 2013, hosted by the institute of interior design of the Architectural Society of China, 6 different regional schools with the disciplines of architecture and design and the major of interior design--Tongji University, South China University of Technology, Harbin Institute of Technology, Xi'an University Of Architecture and Technology , Beijing University of Civil Engineering and Architecture and Nanjing Arts Institute and the famous design companies co-founded the new platform of the CIID "interior design 6+1" school and enterprise joint graduation design to explore the road of cultivating indoor architect under the goal of "excellence program". The 6+1 unit, 6+1 months, 6+1 parts and 6+1 links fit into the number of the CIID "interior design 6+1" school and enterprise joint graduation design.

The CIID "interior design 6+1" activity of school and enterprise joint graduation design has been set up 3 years since it is founded. Every topic of graduation design is guiding by the hot issue which is paid much attention by design industry and society. 2013 (the first session) school and enterprise graduation design named "the business opportunities of post- matches-the reconstruction interior design of the National Stadium", which focused on the management and use of big gymnasiums after large-scale sports matches as an issue paid attention by the society and badly needs to be solved. 2014(the second session) school and enterprise graduation design named "the volume of the new environment of city track--design of Shanghai subway rebuilding", which discussed the issue of city renewal and the environmental design of subway stations.2015 (the third session) school and enterprise graduation design named "the volume of the ordnance heritage- environmental design on Nanjing Chenguang 1865 creative industry park", which talk about the issue of protection and utilization of modern outstanding industrial heritage of historical and cultural city.

《Notification on Enhance the work of cultural heritage protection by state council》which is promulgated by the state council on December 22,2005.There is a strong action on enhance the work of cultural heritage protection in this notification. And also it decided that every second Saturday of June of each year would be "national cultural heritage day" since 2006. June 13, 2015 will be the tenth "national cultural heritage day" in our country. The topic of "national cultural heritage day" in this year is "protecting the cultural heritage, so everyone could enjoy it"

The international committee for the conservation of the industrial heritage (TICCIH) had passed a Nizhny Tagil Charter during July 3 to 10, 2003. This charter states the industrial heritage definition which reflected the basic definition of industrial heritage over the international society. Today, the international societies continual to arouse public realize the importance of understanding the definition of cultural heritage and assessment on the value of cultural heritage in different view and different ways.

The public realized that the conservation of the industrial heritage should be regarded as an indivisible part in cultural heritage in general sense. The conservation of the industrial heritage is inheritance of human culture and foundation of society culture. The conservation of the industrial heritage also maintains diversity and creativity of culture, which promote the whole society moving forward. The conservation of the industrial heritage contains history value, social value and economic value.

Nowadays, the international societies already form a good atmosphere for the conservation of the industrial heritage. More and more country begins to pay much attention on the conservation of the industrial heritage. Under the rule of preservation planning, a large number of countries promote people to understand more about industrial heritage knowledge through proper utilization and protection of industrial heritage. And this action also enhance transition on regional industry, strengthening environmental management, restore regional competition and attraction, and give us a lot of experience on reviving in national economy and in society.

Nanjing Chenguang 1865 Creativity Industrial Park's predecessor was Jin ling machinery bureau, which was built during the Westernization Movement in late Qing Dynasty, when Li Hongzhang was Governor General of Jiangnan and Jiangxi Provinces in 1865, and it is the cradle of China's military industries and weapons industries. The park covers 210,000 m2, which has a total construction area of 110,000 m2, which is full of historical and cultural advantages, Protective Exploitation of Industrial Heritage. The following is the construction of the park position and goals: cultural creativity and technological innovation as the subject matter of the park to build a well-known financial culture, creativity, science and technology, tourism as an integrated fashion creative cultural industry base.

This activity is in continual to carry out teaching of graduation design which relies on design industry and enterprise. It reflects a phenomenon that the education services in cultivating enterprise talents, which shows some characteristics and abilities and gets a great influence in the relating principles and fields. The Nanjing historical architecture research institute was regarded as the general director unit for the naming and joint graduation design. The process of opening ceremony, the judge and award of the graduation defense was supported strongly by Nanjing Chenguang 1865 real estate investment management co., LTD.

The book The volume of ordnance heritage-the Environmental design of Nanjing chenguang 1865 creative industry park--CIID interior design 6+1 "2015 (the third session) school and enterprise graduation design was published by the China Water Power Press. This book uses the form

前言
探讨工业遗产的保护与利用
创新室内设计人才教育培养

陈静勇
执行主编
中国建筑学会室内设计分会副理事长
北京建筑大学 教授

Chen Jingyong
Executive Editor
Vice President of CIID
Professor of Beijing University
of Civil Engineering
and Architecture

在建设创新型国家和人才强国战略的指引下，立足我国学科专业的新环境，面对室内建筑师教育培养的新目标，2013年，由中国建筑学会室内设计分会主办，由同济大学、华南理工大学、哈尔滨工业大学、西安建筑科技大学、北京建筑大学、南京艺术学院等6所不同地域的、设置建筑学和设计学等学科室内设计方向的高校与知名设计企业，共同创建了CIID"室内设计6+1"校企联合毕业设计活动新平台，探索"卓越计划"目标下的室内建筑师教育培养之路。6+1个单位、6+1个月、6+1个环节、6+1个板块等也形成了CIID"室内设计6+1"校企联合毕业设计活动形象传达的贴切数字。

CIID"室内设计6+1"校企联合毕业设计活动创办3年以来，毕业设计选题始终是以行业、社会密切关注的热点问题为导向。2013（首届）的"赛后商机——国家体育场赛后改造室内设计"课题，涉足了大型体育赛事之后大型体育场馆经营利用的问题；2014（第二届）的"城轨新境——上海地铁改造环境设计"课题，探讨了城市更新与地铁站点的环境设计的问题；2015（第三届）"兵工遗产——南京晨光1865创意产业园环境设计"研习了历史文化名城近代工业遗产保护与利用问题。

2005年12月22日，国务院发布《国务院关于加强文化遗产保护工作的通知》，要求进一步加强文化遗产保护工作。其中一项重要举措就是：决定从2006年起，每年6月的第二个星期六为中国的"文化遗产日"。2015年6月13日是我国的第10个文化遗产日，主题是"保护成果 全民共享"。

国际工业遗产保护联合会（TICCIH）于2003年7月3日–10日通过的《下塔吉尔宪章》中阐述的工业遗产定义反映了国际社会关于工业遗产的基本概念。今天，国际社会正在不断地鼓励多样化地理解文化遗产的概念和评价文化遗产价值的重要性。人们认识到，应将工业遗产视作普遍意义上的文化遗产中不可分割的一部分。保护工业遗产就是保持人类文化的传承，培植社会文化的根基，维护文化的多样性和创造性，促进社会不断向前发展。工业遗产具有重要历史价值、社会价值、经济价值等。目前，国际社会对于工业遗产保护逐渐形成良好氛围，越来越多的国家开始重视保护工业遗产，在制定保护规划的基础上，通过合理利用使工业遗产的重要性得以最大限度的保存和再现，增强公众对工业遗产的认识。在推动地区产业转型、积极整治环境、重塑地区竞争力和吸引力、带动经济社会复苏等方面，取得了不少成功的经验。

南京晨光1865创意产业园，其前身是清末洋务运动期间，时任两江总督的李鸿章于1865年创建的金陵机器局，是中国军事工业和兵器工业的摇篮。园区占地面积21万平方米，总建筑面积11万平方米。园区充分发挥历史文化优势，保护性开发利用工业建筑遗产。园区建设定位和目标是：以文化创意和科技创新为主题内容将园区打造成为国内外知名的融文化、创意、科技、旅游为一体的综合性时尚创意文化产业基地。

本届活动继续坚持依托行（企）业开展联合毕业设计教学，体现教育服务于人才培养需求的原则。经学会遴选，委托南京观筑历史建筑文化研究院负责本届联合毕业设计活动命题和总指导工作，开题踏勘和过程评审等得到了南京晨光1865置业投资管理有限公司的大力支持。《兵工遗产卷：南京晨光1865创意产业园环境设计——CIID"室内设计6+1"2015（第三届）校企联合毕业设计》由中国水利水电出版社出版；全书采用中英文对照方式，记载了活动的命题研讨、开题踏勘、中期检查、答辩评审与表彰奖励、编辑出版、专题展览、对外交流等的过程印记。

本届活动的前4个教学环节，从2014年11月5日（厦门市，学会二十四届年会，厦门大学）单独召开CIID"6+1室内设计"校企联合毕业设计研讨会开始，经历2015年3月7日–8日（南京市，南京艺术学院）开题仪式和南京晨光1865创意产业园现场踏勘、4月18日–19日（西安市，西安建筑科技大学）中期检查，到6月6日–7日（广州市，华南理工大学）答辩评审会和表彰奖励仪式结束，学会、学校、企业、专家等多方协同，就南京晨光1865创意产业园的工业遗产保护与室内设计、工业遗产保护与

of Chinese-English to records the process of research of the proposition, opening ceremony and site survey, the mid-term defense, the judge and award of the graduation defense, editing and publishing, exhibitions, external exchanges.

The first 4 teaching sections began with the Seminar of "the CIID interior design 6+1 school and enterprise joint graduation design" of November 5,2014(Xiamen, the 24th annual meeting of the Architectural Society of China, Xiamen university), and went through the opening ceremony (Nanjing, Nanjing Arts Institute) and the site survey of Nanjing chenguang 1865 creative industry park from March 7 to March 8, 2015, the mid-term examination (Xi'an, Xian University of Architecture and Technology) from April 18 to 19, and ended with the judge and award of the graduation defense (Guangzhou, South China University of Technology). The academy, universities and enterprise jointly carried out a comprehensive practice teaching activity on the 3 design project about the conservation of the industrial heritage and interior design, the conservation of the industrial heritage and exhibition design, the conservation of the industrial heritage and landscape design of Nanjing chenguang 1865 creative industry park. The teachers and students had a true feeling about the conservation and utilization of the industrial heritage, which made a promotion for the teaching level of graduation design.

The last two sections of this session is also going well according to the plan. The volume of the ordnance heritage: CIID "Interior Design 6+1" of 2015(the third session) of school and enterprise joint graduation design is editing by the Architectural Society of China and Beijing University of Civil Engineering and Architecture. The drafts filled with the cooperation achievement of the Architectural Society of China, the universities, and the experts have already been collected to Beijing. The theme exhibition will be showed on the 25th annual conference and international academic exchange (Ganqing) from October 19 to 20.

The volume of the ordnance heritage uses "6+1 Activity steps"(the process of this activity) as editing style. The basic teaching materials of the joint graduation such as the Constitution (2015 revised edition), the outline, the task book of the graduation framework, the rules of the judge and award of defense are summarized in the "general rule of the activity", which is convenient for looking up in the future.

The three new chapters of this book are "site survey", "process unit", "research article", which is used as the conclusion of promoting process of teaching and achievement of study in this activity.

According to the theme of 3"thematic design" stations, we should collect and edit the work of each school, make a contrast of the design concepts, highlight the emphasis of the problems and attach the comments of the experts and directors, the experience of the students, award certificates and the photos of the teachers and students. Further, the "expert advice" station could be arranged. In the end, the videos and photos reflecting each link of the activity could be fixed in the "activity memory" station.

This book is determined to recommend as teaching reference book of CIID.

The volume of the ordnance heritage will be released at the same time of the Ganqing annual conference. It also will be expended section as a distinctive example of conducting interior design education exchange with the AIDIA and the IFI.

In a word, because of the large number of the activity memory wanted to edit and the limit of the space of the article, there may be unavoidable careless omissions in editing. Please finger out and we will make it better.

Thank you for the direction and support of the committee of national institutions of higher education architecture discipline guiding, the sub-committee of higher education arts & design major instructor guiding of MOE, the sub-committee of higher education industrial design major instructor guiding of MOE.

Thank you for the help and direction of the related local professional committees of the Architectural Society of China and the colleges and universities participated in this activity, the Nanjing historical architectural culture research institute, Nanjing Chenguang 1865 real estate investment management co., LTD and the lecture and judging lecture experts.

Thank you for the good preparations of the host schools for this session and the volunteers

Thank you for the strong support of china water power press and the hard work of the editor

The CIID "Interior Design 6+1" activity of 2015(the third session) school and enterprise joint graduation design ended with a big success. The graduates from the 4 majors attended the graduation ceremony and degree awarding ceremony of their own schools. Among them, architectural graduates received bachelor of architecture, industrial design graduates received bachelor of engineering, environmental design (the original art design) and art and Technology (the original exhibition art and Technology) graduates received Bachelor of Arts. The last summer vacation during the university became a new start for graduates to step on the way to meet the need of their work and to be professional.

This year is the 26th year of the CIID since it is founded. It is a thing be worth of celebrating for the Architectural Society of China to use the volume of the ordnance heritage as the innovative achievement of interior design education and present it to all the friends.

To cultivate a large number of interior architects with a strong innovative ability who can adapt to national innovation driven development strategy and serve the needs of social development is a proposition for the authorities of the industry sectors, associations, schools, enterprises and society to face together.

July 23, 2015

展示设计、工业遗产保护与景观设计等3个"设计专题",联合开展了综合性实践教学活动。师生们对历史文化名城近代工业遗产保护与利用问题的协同探讨感同身受,交流深入,促进了毕业设计教学水平的提升。

本届活动的后两个展示环节也在按计划进行。《兵工遗产卷》由学会和北京建筑大学联合主编,满载着学会、高校、企业、专家等协作成果的书稿从东西南北中陆续汇集起来;专题展览将于10月19-20日在学会2015年二十五届(甘青)年会暨国际学术交流会上呈现。

《兵工遗产卷》以6+1个"活动脚步"板块作为编辑体例,将活动《章程》(2015修订版)及本届活动《纲要》《毕业设计框架任务书》《答辩活动准备要求与评审、奖励规则》等联合毕业设计教学基础资料归集在"活动总则",以利今后查阅;新增"调研踏勘""过程方案""教研论文"等3个板块内容,以增强对教学过程、教学研究成果等的总结;按3个"专题设计"分别汇编各校作品,列比作品设计概念,突出解决问题侧重点,附上专家点评、学生感言、获奖证书、师生照片等;进而安排"专家寄语";最后将反映活动各环节的影像定格在"活动印记";新确定了将该书列作中国建筑学会室内设计分会推荐的教学参考书。

《兵工遗产卷》将于(甘青)年会同期出版发行,也作为与亚洲室内设计联合会(AIDIA)和国际室内建筑师/设计师联盟(IFI)等开展室内设计教育交流的特色案例,成为活动的一个拓展环节。

总之,希望编入的活动印记较多,限于篇幅,编辑之中难免挂一漏万,敬请指正,以利改进。

感谢全国高等学校建筑学学科专业指导委员会、教育部高等学校艺术设计类专业教学指导分委员会、教育部高等学校工业设计专业教学指导分委员会等长期以来对高校相关学科专业建设工作的指导!

感谢学会相关地方专业委员会、参加高校、南京观筑历史建筑文化研究院、南京晨光1865置业投资管理有限公司、专题讲座与评审专家等对本届活动的指导和帮助!

感谢本届活动承办高校的悉心筹备,以及甘当志愿者的朋友们!

感谢中国水利水电出版社的大力支持和编辑们的辛勤工作!

CIID"室内设计6+1"2015(第三届)校企联合毕业设计活动已圆满结束,从工业遗产保护与室内设计、展示设计、景观设计等协同设计教学活动中走过来的建筑学、工业设计、环境设计、艺术与科技4个专业的毕业生们,已成为了新一届的建筑学学士(专业学位)、工学学士、艺术学学士,他们即将踏上面向行业需求、走向卓越的新征途。

今年是中国建筑学会室内设计分会成立26周年。在此,以《兵工遗产卷》作为学会室内设计教育工作创新成果之一,呈献给朋友们。响应国家创新驱动发展战略和服务经济社会发展需求,培养造就一大批创新能力强、适应需要的高质量室内建筑师,始终是行业主管部门、学(协)会、高校、企业、社会等共同面对的命题。

二〇一五年七月二十三日

联合毕业设计开题活动（2015/03/07 南京）

CIID"室内设计6+1"2015（第三届）校企联合毕业设计 全体师生合影
拍摄于南京艺术学院

Preface 序 005	**Industrial Heritage Protection and Interior Design** 04 工业遗产保护与室内设计 064
Forword 前言 007	Jinling Seal 金陵印 066
	MATRIX 基质的再生 074
Introduction of the Activity 01 活动总则 014	Blending 交融 082
Activity Outline 2015 届活动纲要 016	Stop Fighting for Forest 止戈为林 090
The Artides of Association 2015 届活动章程 018	Growth and Carsh 生长和碰撞 100
The Framework of the Task 2015 届框架任务书 020	Introducing the Flowing Water to Interior Design 渠水流方 106
Report of Building Plan 附建筑图 022	Wind Sing of Jinling 金陵风吟 112
Defense Rules 答辩规则 024	Old Machine on the Wall, Old Handcraft under the Wall 城墙上的老机器，城墙下的老手艺 120
	Culture Worming Machine——1865 Flim Factory 文化蛹动机——1865 影工厂 126
Reconnaissance Survey 02 调研踏勘 026	Feeling · Space 时 · 境 132
Process Plan 03 过程方案 042	Zigzag Jinling——From Manufacturing to Intelligent Production 矩象金陵——从制造到智造 136

Li · Fire
离火 144

Industrial Heritage Protection and Landscape Design
05 工业遗产保护与景观设计 **152**

45° Scene
45°光景 154

Stop fighting for Forest
止戈为林 162

Growth and Carsh
生长和碰撞 169

Convergence · Generation
融·生 174

Passing through Jinling 2065
穿越金陵 2065 182

Industrial Heritage Protection and Display Design
06 工业遗产保护与展示设计 **188**

Culture Worming Machine——1865 Flim Factory
文化蛹动机——1865 影工厂 190

Machine Gun Museum
机枪博物馆 199

Introducing the flowing water to interior design
渠水流方 203

Wind Sing of Jinling
金陵风吟 208

Old Machine on the Wall, Old Handcraft under the Wall
城墙上的老机器，城墙下的老手艺 213

Reconstruction of the Integrated Service Area
A2 综合服务区改造 216

Teaching Graduation Design
07 联合毕业设计的教学探讨 **220**

Three Relations and Three Principles: Teaching exploration of third sessions of "Interior design 6+1" University and Enterprise Joint in Graduation Design
三个关系和三个准则
——三届"室内设计 6+1"校企联合毕业设计的教学探索
左琰 222

Reflection on the "Interior Design 6+1" University and Enterprise Joint in Graduation Design
"室内设计 6+1"校企联合毕业设计活动随想
姜文艺 陈建华 226

Brief Discussion on Renovation of Consumption Space of Old Buildings
浅谈旧建筑改造消费空间的启示与思考
马辉 刘杰 周立军 230

Exploration and Practice of Teaching Research of School and Enterprise Joint Graduation Design
校企联合毕业设计教学研究探索与实践
刘晓军 何方瑶 234

Practice and thinking of diversified University Enterprise Joint Graduation Design: Insights on "Interior Design 6+1" 2015 University Enterprise Joint Graduation Design
多元化校企联合毕业设计实践与思考
——2015 年"室内设计 6+1"校企联合毕业设计感悟
朱宁克 杨琳 238

Research on the Current Situation of Exhibition Design in China
我国展示设计现状研究
朱飞 242

Discussion on the "whole-process node-control" teaching design to graduation design
对"全过程节点把控"式毕业设计教学设计的探讨
吕勤智 黄焱 246

Activity Mark
08 活动印记 **250**

Message
寄语 252

Tidbits
花絮 260

Unit Introduction
单位介绍 262

附

Acknowledgement
感谢 272

CIID"室内设计 6+1"2015（第三届）校企联合毕业设计
CIID "Interior Design 6+1" 2015(Third Scssion)University and Enterprise Joint in Graduation Design

活动总则
Introduction of the Activity

01

2015 届活动纲要

一、课　　题：兵工遗产——南京晨光 1865 创意产业园环境设计

二、项目地点：江苏省南京市秦淮区应天大街 388 号；南京晨光 1865 创意产业园 A-1、A-2

三、主办单位：中国建筑学会室内设计分会（CIID）

四、承办高校：CIID 二十四届（厦门）年会（命题研讨）
南京艺术学院（开题仪式、现场踏勘）
西安建筑科技大学（中期检查）
华南理工大学（答辩评审、表彰奖励）
北京建筑大学（《兵工遗产卷》总编）

五、参加高校（学院\专业）：
同济大学（建筑与城市规划学院\建筑学）
华南理工大学（建筑学院\建筑学）
哈尔滨工业大学（建筑学院\环境设计）
西安建筑科技大学（艺术学院\环境设计）
北京建筑大学（建筑与城市规划学院\工业设计）
南京艺术学院（工业设计学院\艺术与科技）
浙江工业大学（艺术学院\环境设计）

六、命题企业：南京观筑历史建筑文化研究院

七、出版企业：中国水利水电出版社

八、媒体支持：中国室内设计网 http://www.ciid.com.cn

九、时　　间：2014/11-2015/10

十、活动安排：

序号	阶段	时间	地点	活动内容	相关工作
1	命题研讨	2014/11/05-2014/11/08	学会二十四届（厦门）年会；厦门大学	● 11/05，报到命题研讨会 ● 11/06，学会二十四届（厦门）年会开幕式及学术交流活动 ● 11/08，年会室内设计教育论坛	命题企业负责起草《2015（第三届）校企联合毕业设计框架任务书》 ● 研讨《框架任务书》 ● 商讨确定承办高校工作
2	教学准备	2014/11/09-2015/03/05	各高校	● 校企联合毕业设计教学工作准备	● 各高校报送参加毕业设计师生名单 ● 各高校反馈对《活动纲要》的修改意见和建议 ● 各高校结合参加专业实际，依据《框架任务书》，分别编制本校该专业《毕业设计详细任务书》 ● 各高校安排文献检索与毕业实习 ● 开题仪式、现场踏勘等活动准备 ● 学会和各高校邀请支持企业 ● 开题仪式和现场踏勘准备
3	开题踏勘	2015/03/06-2015/03/08	南京艺术学院	● 03/06，报到 ● 03/07，开题仪式专题讲座 ● 03/08，现场踏勘调研交流	● 举行开题仪式 ● 学会安排专题讲座；颁发讲座专家聘书 ● 安排现场踏勘，核对图纸、补充测绘、拍照、访谈等 ● 编制调研 PPT，进行调研工作交流 ● 商议中期检查、编辑出版等相关工作

4	方案设计	2015/03/09-2015/04/16	各高校	• 方案设计	• 各高校安排相关讲授、辅导、设计、研讨等 • 选择设计专题，明确方案设计目标，完成相关文案、图表等 • 完成调研报告、方案设计基本图示、效果图、模型、分析图表及等相应成果 • 中期检查准备；编制中期检查方案设计汇报 PPT
5	中期检查	2015/04/17-2015/04/19	西安建筑科技大学	• 04/17，报到 • 04/18，中期检查 • 04/19，专题讲座　　参观实习	• 中期检查汇报、专家点评 • 研讨方案设计深化重点 • 商议答辩评审、表彰奖励、编辑出版、展览等相关工作
6	深化设计	2015/04/20-2015/06/04	各高校	• 方案深化设计	• 完成方案深化设计图示、效果图、模型、分析图表、设计说明等相应成果，制做展板（A0竖版,3张/方案组），提交电子版 • 编制毕业设计答辩汇报 PPT，提交电子版 • 《兵工遗产卷》分配页面排版（每方案6P）；提交电子版 • 答辩评审、表彰奖励准备
7	答辩评审	2015/06/05-2015/06/07	华南理工大学	• 06/05，报到 • 06/06，答辩、评审 • 06/07，表彰奖励	• 答辩布展与观摩 • 毕业答辩；等级奖评审 • 表彰毕业设计等级奖、优秀毕业设计指导教师、优秀毕业设计组织单位、校企联合毕业设计突出贡献企业等 • 学会室内设计教育论坛策划和2016（第四届）校企联合毕业设计命题初步酝酿
8	编辑出版	2015/06/08-2015/09/30	学会、参加高校、命题企业、支持企业、出版企业	• 07/10前，完成《兵工遗产卷》分学校部分书稿编辑工作 • 08/15前，完成书稿总编工作 • 08/30前，完成书稿校审工作，形成清样，送印厂 • 09/30前，《兵工遗产卷》出版	• 各高校和相关专家提交《兵工遗产卷》书稿 • 学会和北京建筑大学负责书稿总编 • 出版企业负责书稿校审 • 学会二十五届（甘青）年会"室内设计6+1"总结研讨会《兵工遗产卷》发行式准备
9	展览交流	2015/10/16-2015/10/20	学会二十五届（甘青）年会室内设计教育论坛	• 10/16，报到 • 10/19，2015（第三届）校企联合毕业设计活动总结与2016（第四届）校企联合毕业设计命题研讨 • 10/18，《兵工遗产卷》发行式 • 10/19-20 学会年会专题展览	• 展板布展与观摩 • 学会二十五届（甘青）年会"室内设计6+1"总结研讨会《兵工遗产卷》发行式 • 学会室内设计教育成果交流活动

The Artides of Association

In order to meet the need of cultivating special talents of interior design in the field of urban and rural construction, strengthen the pertinence of indoor architect cultivation, promote the relevant institutions of higher education's major construction on interior design discipline and communications on teaching and guide the relevant majors conduct teaching work on graduation design towards architecture, the institute of interior design of Architectural Society of China(CIID, Association for short in the following) acted as the director of advocacy and the colleges and universities who have interior design and the relating disciplines and well-known architectural and interior design enterprises carry out joint graduation design together.

In order to make the graduation design to be normative and to form its brand and characteristics, the CIID made regulations based on the opinions and advices of the relevant schools. The regulations passed on the CIID "interior design 6+1" 2013 (first) school and enterprise joint graduation design proposition meeting and are announced now.

Ⅰ.The background, aim and significance of the school and enterprise joint graduation design

In 2010, the Ministry of Education launched the "Excellent Engineer education program", and announced the subject and major list added to the plan of the "education program for excellent Engineer" from 2011 to 2013. The Committee of degree of the State Council and the Ministry of education publish the "discipline catalogue of degree awarding and talent cultivating (2011)" and added the discipline of "Arts (13)", making "Design Science (1305)" as the first class discipline under Arts. The environmental design is advised as the second class discipline under design science and the interior design is advised as the second class discipline under design science and the interior design is advised as the second class discipline under Architecture. In 2012, the Ministry of Education announced the "catalog of undergraduate majors of colleges and universities" (2012), setting the major of "Design Science (1305)" under" Arts science" and making "environment design" (130503) as its core majors. The independent setting of the "Arts", the setting of the first class principle of design science and the setting and adjusting of interior design and other majors have formed the new professional pattern of the environment design education and the cultivating of special talents of interior design.

The holding of the interior design school and enterprise joint graduation design has a great effect on strengthening the characteristic of relevant disciplines, deepening the teaching exchanges of the graduation design, promoting the innovation of interior design education and cultivating the interior design special talents needed by the service industry during the conduction of the "Excellent Engineer education program" held by the ministry of education.

Ⅱ.The organizations of the school and enterprise joint graduation design

1. The host unit

School and enterprise joint graduation design activity was hosted by the CIID (or the authorized local professional committee), and obtained the guidance and support of the professional high school science teaching guidance committee of architecture, the Ministry of education teaching and guidance committee of design and the Ministry of education teaching and guidance committee of industry design.

2. The participant and host schools

Usually, the 6 universities and colleges with the near discipline and specialty conditions and the relevant majors of interior design became the participant schools through consulting and organizing. There was some area span between the locations of the participant schools, and formed a certain interdisciplinary and collaborative design conditions in subject between.

Every year, there will be one to four schools be chosen as the host schools for the opening ceremony and site survey, the mid-term defense, the judge and award of the graduation defense and the editing and publishing of the school and enterprise joint graduation design of that session.

The proper number of students participating in the joint graduation design for each school is six and the students must be companied with one to two directors. Among the teachers, there must be at least one has senior professional title and be familiar with environment design, interior design and other industry practice business and has a wide communication with the relevant enterprises.

3. The proposition enterprise

The schools hosting the school and enterprise joint graduation design opening ceremony and site survey are responsible for recommending the CIID (or the authorized local professional committee) one famous architecture and interior design enterprise in their provinces (cities) to be as the proposition enterprise. The proposition enterprise are responsible for creating the proposition, making the sketch task book of the graduation design, participating in the opening ceremony and site survey, mid-term examination, the judge and award of the graduation defense, editing and publishing, exhibitions, external exchanges.

4. The supporting enterprise

The participant schools are responsible for recommending the CIID (or the authorized local professional committee) one famous architecture and interior design enterprise in their provinces (cities) to be as the supporting enterprise during every session of joint graduation design. The CIID (or the authorized local professional committee) and the supporting enterprises should sign the agreement of activity supporting and feedback and arrange the supporting enterprise to participate in the graduation design. The CIID (or the authorized local professional committee) is responsible for communicating the selected experts of the supporting enterprise, giving special lecture around the topic of the graduation design, participating in the in the opening ceremony and site survey, mid-term examination, the judge and award of the graduation defense, editing and publishing, exhibitions, external exchanges.

5. The pressing enterprises

Based on every session of school and enterprise joint graduation design, the CIID (or the authorized local professional committee) should select one famous pressing enterprise to take the responsibility of the publishing work of the school and enterprise joint graduation design.

Ⅲ.The process of the school and enterprise joint graduation design

1. According the graduation teaching work of the participant schools, the school and enterprise joint graduation design is held once every year (according to the session of the graduates).

2. The sections of the school and enterprise joint graduation design mainly include the researching of the topic of the graduation design, the opening ceremony and site survey, mid-term examination, the judge and award of the graduation defense, editing and publishing of the school and enterprise joint graduation design of that session and exhibitions. It usually takes 6 months for the teaching section of the school and enterprise joint graduation design. External exchange is regarded as one extending section of the school and enterprise joint graduation design.

3. The CIID (or the authorized local professional committee), as the host unit of this activity, is responsible for the overall planning, publicity, coordinating the participant schools, the relevant enterprises and the exhibition institutions, employing the relevant experts to hold special academic lectures, organizing annual award and the selection and awarding of the excellent guiding teachers, organizing units and special contribution of graduation design as well as the education and international exchange of interior design.

4. The participant schools, together with the relevant enterprises, should draw out the outline of the activity of the school and enterprise joint graduation design under the guidance of the CIID (or the authorized local professional committee) and then the outline is sent to the CIID (or the authorized local professional committee) for approving. The host schools should finish the work of the opening ceremony of the joint graduation and site survey, mid-term examination, the judge and award of the graduation defense, editing and publishing, exhibitions.

5. The research of the proposition

The guiding teacher of the participant schools should attend the research meeting of the proposition of the joint graduation design. The graduation design topic was formed by the proposition enterprise and participant schools by focusing on city design, architectural design, environmental design, the frontier of the interior design development and hot issues in industry and combining with the teaching reality of graduation design of the participant schools. Then the topic is sent to the CIID (or the authorized local professional committee) for approving. The graduation design proposition is required to have the approaches to collecting some relevant design data and site survey. The proposition research is generally arranged in the fall semester of college (around November each year), and be held with the education forum of the academic annual meeting of the CIID (or arrangement of seminars).

6. The opening ceremony and site survey

The graduate teachers and students should attend the opening ceremony and site survey of the joint graduation design. The proposition enterprise provides the essential basic materials and task of the graduation design and arranges site survey and other activities. The opening activity is generally arranged in the spring semester of college (around the first ten days of March).

7. The mid-term examination

The graduate teachers and students should attend the mid-term examination of the joint graduation design. The guiding teacher of the participant schools of the joint graduation design and the experts of the relevant enterprises should check and review the mid-term results of the graduation design and conduct teaching exchanges. The mid-term examination is generally arranged in the spring semester of college (in the last ten days of April).

During the mid-term examination, each university can choose not more than 3 defense project to have the statement and achievements exhibition; each graduation of the program group should choose their designing theme for the defense. Each statement time is 10 minutes; question-and-answer is not more than 10 minutes.

8. The judge and award of the graduation defense

The graduate teachers and students should attend the judge and award of the graduation defense. The experts of the CIID (or the authorized local professional committee), guiding teacher of the participant schools of the joint graduation design and the experts of the relevant enterprises should attend the judge and award of the graduation defense. The judge and award of the graduation defense is generally arranged in the end of the spring semester of college (in the first ten days of June).

During the graduation defense, each university can choose not more than 2 defense project to have the statement and achievements exhibition; each graduation of the program group should choose their designing theme for the defense. Each statement time is 15 minutes; question-and-answer is not more than 15 minutes. In addition, each university can choose not more than 2 recommendation schemes to have the achievements exhibition, which can not have the statement and defense.

On the base of defense and judge, the CIID (or the authorized local professional committee) organizes the selection of the annual award, the excellent guiding teachers, the excellent organizing units and special contribution of graduation design and so on and gives awards to them(The proportion of awards can be set at 1:2:4). The graduation design annual award is set according to the design project and the level award is set.

2015届活动章程

为服务城乡建设领域室内设计专门人才培养需求，加强室内建筑师培养的针对性，促进相关高等学校在室内设计学科专业建设和教育教学方面的交流，引导相关专业面向建筑行（企）业需求组织开展毕业设计教学工作，由中国建筑学会室内设计分会（CIID，以下简称学会）倡导、主管，国内设置室内设计相关学科专业的高校与知名建筑与室内设计企业开展联合毕业设计。

为使联合毕业设计活动规范、有序，形成活动品牌和特色，学会在征求相关高等学校意见和建议的基础上形成本章程，并于学会（CIID）"室内设计6+1" 2013（首届）校企联合毕业设计命题会上审议通过，公布试行。

一、校企联合毕业设计活动设立的背景、目的和意义

2010年教育部启动了"卓越工程师教育培养计划"，于2011-2013年分三批公布了进入"卓越计划"的本科专业和研究生层次学科。2011年国务院学位委员会、教育部公布了《学位授予和人才培养学科目录（2011年）》，增设了"艺术学（13）"学科门类，将"设计学（1305）"设置为"艺术学"学科门类中的一级学科。"环境设计"建议作为"设计学"一级学科下的二级学科，"室内设计"建议作为新调整的"建筑学（0813）"一级学科下的二级学科。2012年教育部公布了《普通高等学校本科专业目录（2012年）》，在"艺术学"学科门类下设"设计学类（1305）"专业，"环境设计（130503）"等成为其下核心专业。"艺术学"门类的独立设置、设计学一级学科以及环境设计、室内设计等学科专业的设置与调整，形成了我国环境设计教育和室内设计专门人才培养新的学科专业格局。

举办室内设计校企联合毕业设计活动，对在教育部"卓越工程师教育培养计划"实施中加强相关学科专业特色建设，深化毕业设计各教学环节交流，促进室内设计教育教学协同创新，培养服务行（企）业需求的室内设计专门人才，具有十分重要的意义。

二、校企联合毕业设计活动组织机构

1. 主办单位

校企联合毕业设计活动由学会（或经授权的地方专业委员会）主办，得到了全国高等学校建筑学学科专业指导委员会、教育部高等学校设计学类专业教学指导委员会、教育部高等学校工业设计专业教学指导分委员会等的指导和支持。

2. 参加高校、承办高校

校企联合毕业设计活动一般由学科专业条件相近，设置室内设计方向的相关专业的6所高校通过协商、组织成为活动参加高校。参加高校间所处地理区域具有一定的距离，在学科专业间形成一定的交叉性和协同设计条件。

每年在参加高校中推选1～4所高校分别作为毕业设计开题仪式与现场踏勘、中期检查、答辩评审与表彰奖励、当届《校企联合毕业设计》（主题卷）编辑出版等活动的承办高校。

每所高校参加联合毕业设计学生一般以6人为宜，要求配备1～2名指导教师，其中至少有1名指导教师具有高级职称，熟悉环境设计、室内设计等工程实践业务，与相关领域企业联系较广泛。

3. 命题企业

承办校企联合毕业设计开题仪式与现场踏勘的高校，负责向学会（或经授权的地方专业委员会）推荐所在省（市）的（1家）行业知名建筑与室内设计企业作为毕业设计命题企业。命题企业负责毕业设计课题命题，编制毕业设计框架任务书，参与开题仪式与现场踏勘、中期检查、答辩评审与表彰奖励、编辑出版、专题展览、对外交流等工作。

4. 支持企业

参加高校在每届校企联合毕业设计活动中，分别向学会（或经授权的专业地方委员会）推荐（1家）行业知名建筑与室内设计企业作为毕业设计支持企业，由学会（或经授权的地方专业委员会）与支持企业签订活动支持与回馈协议，安排支持企业参与毕业设计活动。学会（或经授权的地方专业委员会）负责联系支持企业选派专家，围绕毕业设计课题进行专题讲座，参与毕业设计开题仪式与现场踏勘、中期检查、答辩评审与表彰奖励、编辑出版、专题展览、对外交流等工作。

5. 出版企业

学会（或经授权的地方专业委员会）就每届校企联合毕业设计活动，遴选（1家）行业知名出版企业，负责承担当届《校企联合毕业设计》（主题卷）的出版工作。

三、校企联合毕业设计活动组织流程

1. 校企联合毕业设计活动按照参加高校毕业设计教学工作安排在每个年度（按毕业生届次）举行1次。

2. 校企联合毕业设计活动主要教学环节包括：毕业设计命题研讨、毕业设计开题仪式与现场踏勘、毕业设计中期检查、毕业设计答辩评审与表彰奖励、当届《校企联合毕业设计》（主题卷）编辑出版、专题展览等（6个）主要环节。校企联合毕业设计的教学环节时间跨度一般为6个月，对外交流作为联合毕业设计活动的（1个）扩展环节。

3. 学会（或经授权的地方专业委员会）作为活动主办单位，负责活动总体策划、宣传，协调参加高校、相关企业、展览机构等，聘请有关专家举办专题学术讲座，组织毕业设计学年奖、毕业设计优秀指导教师、毕业设计优秀组织单位、毕业设计特殊贡献奖等的评选、表彰，以及室内设计教育国际交流工作。

4. 参加高校在学会（或经授权的地方专业委员会）的指导、协调下，联合相关企业等，共同拟定校企联合毕业设计活动纲要，报学会（或经授权的地方专业委员会）审定。联合毕业设计开题仪式与现场踏勘、中期检查、答辩评审与表彰奖励、《校企联合毕业设计》（主题卷）编辑出版等工作由承办高校分别落实。

5. 命题研讨

参加高校的毕业设计指导教师参加联合毕业设计命题研讨会。毕业设计课题由命题企业与参加高校，着眼城市设计、建筑设计、环境设计、室内设计等领域发展前沿和行业热点问题，结合参加高校毕业设计教学实际商讨形成，报学会（或经授权的地方专业委员会）审定。毕业设计命题要求具备相关设计资料收集和现场踏勘等条件。命题研讨会一般安排在高校秋季学期中（每年11月左右），结合当年学会学术年会的教育论坛（或安排专题研讨会）进行。

5. 开题仪式与现场踏勘

参加高校的毕业设计师生参加联合毕业设计开题仪式和现场踏勘活动，命题企业提供毕业设计必要的设计基础资料和设计任务需求，安排现场踏勘等活动等。开题活动一般安排在高校春季学期开学初（3月上旬）进行。

6. 中期检查

参加高校的毕业设计师生参加联合毕业设计中期检查活动。参加高校毕业设计指导教师和相关企业专家等，对毕业设计中期成果进行检查、评审，开展教学交流。中期检查一般安排在春季学期期中（4月下旬）进行。

中期检查活动中，每所参加高校优选不超过3个方案组进行陈述与答辩；其中陈述不超过10分钟，问答不超过10分钟。

7. 答辩评审与表彰奖励

参加高校的毕业设计师生参加联合毕业设计答辩评审与表彰奖励活动。学会（或经授权的地方专业委员会）专家、命题企业专家、参加高校毕业设计指导教师等作为评委，参加毕业设计答辩评审、表彰奖励等工作。毕业设计答辩评审与表彰奖励一般安排在春季学期期末（6月上旬）进行。

毕业答辩活动中，每所参加高校优选不超过2个答辩方案组进行陈述与答辩、成果展出；其中陈述不超过15分钟，问答不超过15分钟。此外，每所参加高校可再安排不超过2个自荐方案组进行成果展示，不参加陈述与答辩。

在答辩、成果展示、评审的基础上，学会（或经授权的地方专业委员会）组织评选毕业设计学年奖、毕业设计优秀指导教师、毕业设计优秀组织单位、毕业设计特殊贡献奖等，并给予表彰奖励。毕业设计学年奖按照等级奖（含一、二、三等奖，按照1:2:4比例设置）、优秀奖分别进行评选；其中，等级奖评选仅针对答辩方案设置，优秀奖针对自荐方案设置。

8. 专题展览

学会（或经授权的地方专业委员会）在每届联合毕业设计结束当年学会学术年会上（或经授权的地方专业委员会活动中）安排（1个）毕业设计学年奖作品专题展览（每年10-11月份）；专题展览结束后，毕业设计学年奖作品可在国内相关高校之间巡回展出。

9. 编辑出版

基于每届联合毕业设计活动，各编辑出版（1部）《校企联合毕业设计》（主题卷），作为学会推荐的室内设计、环境设计专业教学参考书。《校企联合毕业设计》（主题卷）编辑工作由学会（或经授权的地方专业委员会）和相应承办高校联合编辑，负责组稿、排版等工作；参加高校作为参编单位，参加高校毕业设计指导教师负责本校排版稿的审稿等工作；出版企业作为责任编辑，负责校审、出版。

活动总则 Introduction of the Activity

9. The exhibitions
The CIID (or the authorized local professional committee) arranges one annual award works thematic exhibition of the graduation design (November-December every year) in the CIID academic annual meeting in the end year of each session of the graduation design (or the authorized local professional committee activities); At the end of the thematic exhibition, the school year awarding works of the graduation can be exhibited in the relevant schools at home abroad.

10. The editing and publishing
Based on the graduation design activities, one volume of "school and enterprise joint graduation design" should be edited per session. The "school and enterprise joint graduation design" is edited by the CIID (or the authorized local professional committee) and the relevant host schools. They are responsible for soliciting and typesetting; the participant schools are as the participating organization. The guiding teachers of the graduation design of the participant schools are responsible for the examination of the manuscripts; as the editor, the publishing enterprise is responsible for proofreading and publishing.

11. External exchange
The "school and enterprise joint graduation design" usually host the issue ceremony in the CIID academic annual meeting in the end year of each session of the graduation design. The CIID carries out the exchange of international interior design by communicating the Asian Interior Design Association (AIDIA) and international interior architect / designer alliance (1F1). In this way, it can open a new door for publicizing the education of Chinese interior design and carrying out international exchange.

IV. The fund of the school and enterprise joint graduation design
1. The CIID (or the authorized local professional committee) is responsible for raising money for the selection of the annual award, the excellent guiding teachers, the excellent organizing units and special contribution of graduation design as well as the thematic exhibition of the annual meeting of the CIID.
2. The participant schools should afford the travel fees of the relevant personnel participating in the school and enterprise joint graduation design and the site fees and relevant fees for the exhibition of the annual awarding works of the graduation design.
3. The host schools afford the site fees, conversation fees, organization fees of the opening ceremony of the joint graduation and site survey, mid-term examination, the judge and award of the graduation defense as well as the editing and pressing fees of the school and enterprise joint graduation design.
4. The proposition enterprise, supporting enterprise and pressing enterprise are responsible for proving the CIID some fund for the school and enterprise joint graduation design activity.

V. Supplementary provisions
1. The constitution was announced and conducted on January 13, 2013 interpreted by the CIID.
2. The constitution was first revised in March. 2014; and second revised in April .2015

The Framework of the Task

The frame task book of the CIID "interior design 6+1"2015(the third session)school and enterprise joint graduation design was brought out by the Nanjing Institute of Architectural Culture View Historical Building, which was entrusted by the CIID and co-determined by the guiding teachers of the school and enterprise joint graduation design of 2015(the third session).The participant schools make the relevant specific task book of the graduation design according the actual teaching work of their own.

The name of the program
Ordnance Heritage——Nanjing Chenguang 1865 Creativity Industrial Park Environmental Design

The location of the program
The predecessor of Nanjing Chenguang 1865 Creativity Industrial Park was Jin ling machinery bureau, which was built during the Westernization Movement in late Qing Dynasty, when Li Hongzhang was Governor General of Jiangnan and Jiangxi Provinces in 1865, and it is the cradle of China's military industries and weapons industries. The park covers 210,000 m2, which has a total construction area of 110,000 m2, which is full of historical and cultural advantages, protective exploitation of industrial heritage. The following is the construction of the park position and goals: cultural creativity and technological innovation as the subject matter of the park to build a well-known financial culture, creativity, science and technology, tourism as an integrated fashion creative cultural industry base.

The range of the design
A1: Built in Republic of China in 23 years (1934), the building is a single layer of reinforced concrete structure, and the building area is 3309.6 m2, whose predecessor is hose company office. The building's length × width × height is 62.1m × 53.1m × 5.77m (steeple is 9.17m);
A2: Built in Republic of China in 23 years (1934), the building is a single layer of reinforced concrete structure, and the building area is 8236.3 m2, whose former predecessor is factory machine room, and the latter is the hose processing section. The building's length × width × height is 92.8m × 93.3m × 5.77m (steeple is 9.17m);
A1, A2 construction site and the outer space surrounding environment.

The subject of the design
Basing on the overall goals that historical city of Nanjing weapon design for conservation and renewal of industrial building heritages for the universities, In a given Nanjing dawn 1865 creative industry garden A1, A2 building and its surrounding space environment, Combining with our school's professional characteristics and the actual that were participated in the activities, they would select to determine the emphasis of the graduation design in the following design project.

1. Architectural Heritage Conservation and interior design.
Comply with the principle of protection of historic buildings; they need to carry out a investigation that combining heritage protection of industrial buildings with interior updated designs. A1 will be transformed into a weapons industry museum and related functional spaces, A2 will be transformed into a creative cultural blocks and a function space; which can be doing architectural heritage conservation planning, zoning update function space, dynamic line design, space environment and facility design, system design knowledge guide.

2. Architectural Heritage Preservation and exhibition design.
Comply with the principle of protection of historic buildings; they need to carry out the exploration that combining the heritage conservation with exhibition design. Based on the military industry and the history of the development of Jinling weapons industry; have a design by means of display design A1, A2 architectural heritage and update functions of the space environment.

3. Architectural Heritage and Landscape Design.
Comply with the principle of protection of historic buildings, they need to carry out the exploration that combining heritage protection of industrial buildings and landscape design. Based on architectural heritage protection planning, A1, A2 external space environment are used to do landscape design, and doing something about public facilities, guiding knowledge systems, public art and other facilities.

the design depth
Design (deepen)

The achievement of the design
(1) Design Notes
Design Notes mainly comprise: elaborate overall design concept, the design derivative process analysis, design, functional analysis, which can be used for illustration.
(2) Drawing
1. General layout, site design, landscape sectional view
2. Building interior floor plan, top view of a cross-sectional elevation
3. Interior and outside environment facilities layout, lead identification system layout, showcasing layout, three views, details
4. Detailing important node
5. The main material tables, furnishings, inventory and other chart
(3) color renderings
1. Interior Design
2. Landscape renderings
3. Public facilities, display props, public art renderings
(4) The outcome of submission
1. Intermediate inspection
Each participant University preferably chooses three program groups to state, each time for a reply is limited to 20 minutes (including the statement of 10 minutes, 10 minutes Q & A); Submit Phase Achievements, submit report stage, and report PPT files.
2. Graduation defense
 (1) Each participant University preferably chooses three program groups to state, each time for a reply is limited to 20 minutes (including the statement of 10 minutes, 10 minutes Q & A); PPT files, work panels, work typesetting is submitted in graduation defense statement;
 (2) Each design group works are limited to three panels, poster format A0 (900 × 1200mm), the resolution is not less than 100dpi. Poster templates are provided unified by the Institute in accordance with the requirements of the exhibition boards. Exhibition is hosted by the University, where the graduation of responsible for defense activities is arranged.
3. Publishing material.
In order to edit and publish"CIID" Interior Design 6 + 1 "2015 (Third) colleges, enterprises and graduation - Ordnance Heritage Volume: Nanjing Chenguang 1865 Creativity Industrial Park Environmental Design", the units and individuals that were related to the activities need to actively respond Society requires, and they are responsible for providing appropriate activities statute, teaching documents, research papers (each instructor wrote one, and less than 2500 words),publishing works (each program 6P), experts (each expert, mentor have to give out a program review within 200 words), experts Message (500 words per expert), student thoughts (200 words or less for each student), working Photos (each expert, teacher, student 1), award certificates (all levels class), Event Highlights (activities of the main links), Introduction of electronic document (1000 words) and other published materials.

Attachment: Building and floor plan
Nanjing Chenguang 1865 Creativity Industrial Park General Plan; A1 and A2 architectural drawings, Photographs before the transformation.

Introduction of the Activity 活动总则

10. 对外交流

《校企联合毕业设计》（主题卷）一般在当届联合毕业设计结束当年学会学术年会上举行发行式，并由学会通过联系亚洲室内设计联合会（AIDIA）、国际室内建筑师/设计师联盟（IFI）等开展国际室内设计教育成果交流，打开（1扇）宣传中国室内设计教育、开展国际交流的新大门。

四、校企联合毕业设计活动经费

1. 学会（或经授权的地方专业委员会）负责筹措评选毕业设计学年奖、毕业设计优秀指导教师、毕业设计优秀组织单位、毕业设计特殊贡献奖等表彰奖励经费，以及学会年会专题展览经费等。

2. 参加高校自筹参加校企联合毕业设计活动相关人员的差旅费，以及毕业设计学年奖作品在本校巡展的场地及相关经费等。

3. 承办高校自筹校企联合毕业设计开题仪式与现场踏勘、中期检查、答辩评审与表彰奖励等相关活动的场地、会议费、组织费，以及《校企联合毕业设计》（主题卷）编辑出版经费等。

4. 命题企业、支持企业、出版企业等负责向学会提供一定的对校企联合毕业设计活动的资助经费等。

五、附则

1. 本章程2013年1日13日通过并公布施行，由中国建筑学会室内设计分会负责解释。

2. 本章程2014年3月第一次修订；2015年4月第二次修订。

2015届框架任务书

CIID"室内设计6+1"2015（第三届）校企联合毕业设计框架任务书由中国建筑学会室内设计分会委托南京观筑历史建筑文化研究院提出，经2015（第三届）校企联合毕业设计指导教师共同商讨确定。活动参加高校应结合本校毕业设计教学工作实际，据此编制相应的毕业设计详细任务书。

一、项目名称
兵工遗产——南京晨光1865创意产业园环境设计

二、项目地点
南京晨光1865创意产业园，其前身是清末洋务运动期间，时任两江总督的李鸿章于1865年创建的金陵机器局，是中国军事工业和兵器工业的摇篮。园区占地面积21万平方米，总建筑面积11万平方米。园区充分发挥历史文化优势，保护性开发利用工业建筑遗产。园区建设定位和目标是：以文化创意和科技创新为主题内容将园区打造成为国内外知名的融文化、创意、科技、旅游为一体的综合性时尚创意文化产业基地。

三、设计范围
A1：始建于民国二十三年（1934年），建筑为单层钢混结构，建筑面积3309.6平方米，原为软管公司工房。建筑物长×宽×高为62.1m×53.1m×5.77m（尖顶9.17m）；

A2：始建于民国二十三年（1934年），建筑为单层钢混结构，建筑面积8236.3平方米，原为12分厂机加工房、后为软管加工工段。建筑物长×宽×高为92.8m×93.3m×5.77m（尖顶9.17m）；

A1、A2建筑外部场地与周边空间环境。

四、设计专题
参加高校基于南京历史城市兵器工业建筑遗产保护与更新设计探讨的总体目标，在给定的南京晨光1865创意产业园中A1、A2建筑及其周边空间环境中，结合本校参加活动专业特色和实际，在以下设计专题中选择确定毕业设计的侧重点：

1. 建筑遗产保护与室内设计

遵从历史建筑保护原则，开展将工业建筑遗产保护与室内更新设计相结合的探讨。将A1改造成兵器工业博物馆及关联功能空间，将A2建筑改造成创意文化街区功能空间；分别进行建筑遗产保护规划、更新功能空间区划、动线设计、空间环境及设施设计、导识系统设计等。

2. 建筑遗产保护与展示设计

遵从历史建筑保护原则，开展将工业建筑遗产保护与展示设计相结合的探讨。基于中国军事工业和兵器工业金陵发展史，借助A1、A2建筑遗产和更新功能空间环境进行展示设计等。

3. 建筑遗产保护与景观设计

遵从历史建筑保护原则，开展将工业建筑遗产保护与景观设计相结合的探讨。基于建筑遗产保护规划，对A1、A2外部空间环境做景观设计，并做公共设施、导识系统、公共艺术等配套。

五、设计深度
方案设计（深化）

六、设计成果

（一）设计说明

设计说明内容主要包含：整体设计思想的阐述、设计衍生过程分析、设计风格、功能分析等，可作图示。

（二）图纸
1. 总平面图、场地设计图、景观断面图
2. 建筑室内平面图、顶面图、剖立面图
3. 建筑室内外环境设施布置图、导识系统布置图、展陈布置图、三视图、详图
4. 重要节点详图
5. 主要材料表、家具陈设清单等图表

（三）彩色效果图
1. 室内效果图
2. 景观效果图
3. 公共设施、展示道具、公共艺术等效果图

（四）成果提交
1. 中期检查

每个参加高校优选3个方案组进行陈述，每组答辩时间限20分钟（含陈述10分钟、问答10分钟）；提交阶段成果汇报PPT文件等。

2. 毕业答辩

（1）每个参加高校优选2个答辩方案组进行陈述，每组答辩时间限30分钟（含陈述15分钟、问答15分钟）；提交毕业设计答辩陈述PPT文件、作品展板、作品排版等。

（2）每个方案设计组的作品展板限3张，展板幅面A0（900×1200mm），分辨率不小于100dpi。展板模板由学会按照年会展板要求统一提供。展览由毕业设计答辩活动承办高校负责布置。

3. 出版素材

为编辑出版《CIID"室内设计6+1"2015（第三届）校企联合毕业设计——兵工遗产卷：南京晨光1865创意产业园环境设计》，活动相关参加单位和个人等需积极响应学会要求，负责提供相应的活动章程、教学文件、教研论文（每所高校指导教师联名写1篇，2500字以内）、排版作品（每个方案6P）、专家点评（每位专家、导师对每个方案点评200字以内）、专家寄语（每位专家500字左右）、学生感想（每个学生200字以内）、工作照片（每位专家、导师、学生1张）、奖励证书（各级各类）、活动花絮（活动各主要环节）、单位简介（1000字以内）等出版素材的电子文档。

七、附建筑与场地图
南京晨光1865创意产业园总平面图、A1与A2建筑图、改造前照片。

附建筑图　Report of Building Plan

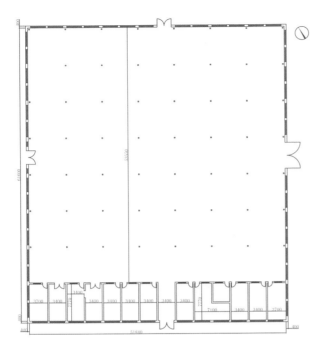

A1 平面图

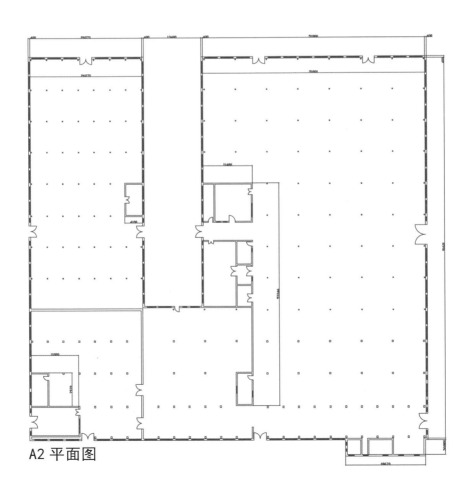

A2 平面图

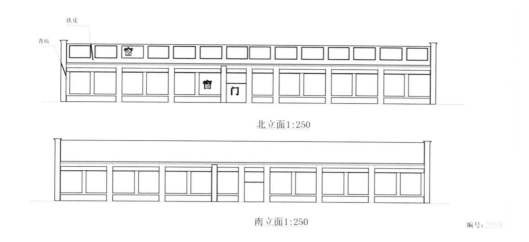

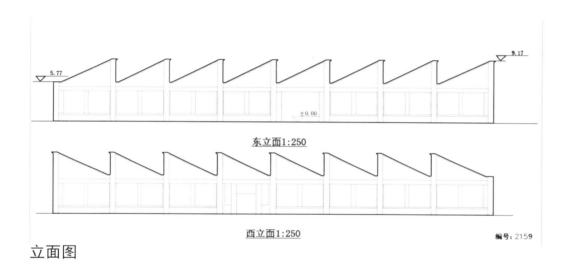

立面图

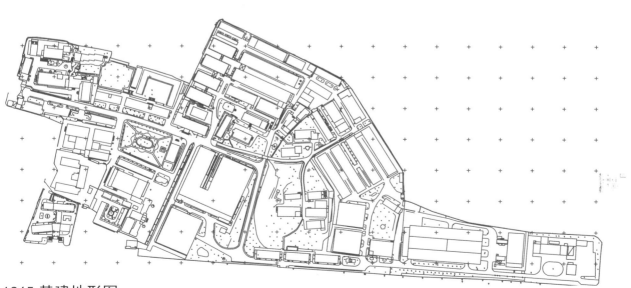

1865 基建地形图

Defense Rules

The preparation requirement of graduation defense of the participant schools and the achievements show
1. The guiding teachers from each school is not more than 2; the number of attending students is limited to six.
2. Every University selects two graduation defense teams, which take part in the graduation design of defense activities. The selection of "Graduation designing grade award" and the achievements show; in additions, we can also apply for two recommendation schemes at most, which take part in the selection of "Graduation designing excellence award and the achievement show", but it can not be take part in the graduation defense and the selection of "Graduation designing grade award".
3. The participant schools prepare the electronic documents of the graduation design defense statement PPT in a project design group and sent them to the Organizing Committee the graduation when checking in for the graduation defense activity.
4. Each design team should prepare 3 design work panels. The panel format is 900*1800 mm, and the resolution is not less than 100 dpi, which uses the same pattern. The electronic version of the display show be sent to the mailbox of the organizing committee a week ahead and the host school for the section of defense award is responsible for the summary printing decoration etc.
5. The design group will edit the manuscript on the basis of the publishing requirements of the CIID "interior design 6+1" school and enterprise joint graduation designment and send them to the Organizing Committee at the scene of the graduation design defense.
6. Graduation design subject direction
Thematic reported election of defense design considering the reality of their own schools, the schools should have a selection to two graduation design defenses, which according to the framework of the CIID Task at last, we organize graduates to have the defense.
The award and judge of the graduation defense
1. The composition of the defense review Group Committee
(1) The CIID (or the authorized local professional committee), proposition enterprise and the supporting enterprise recommend 5-7 experts to be as the out-of-school judges.
(2) The participant schools recommend one guiding teacher to be the school judge of the graduation design.
2. The selection of the "Graduation designing grade award"

(1)"Graduation Award" sets first, second and third prize on the basis of the graduation design of the framework of "interior design 6+1" school-enterprise joint graduation design; The proportion of awards can be set at 1:2:4;Level vacancies, the vacancy number of awards shall be transferred between adjacent levels.
(2)First, by means of the defense, the out-of-school judges and the school judges shall suggest the "Graduation designing grade Award" program Sort. Each graduation of the program group should choose their designing theme for the defense. Each statement time is 15 minutes; question-and-answer is not more than 15 minutes. Two judges shall fill out the ballot, and sort the selection (for example, 1 is the recommended sort first, 2 is the recommended second, repeat).Organizing Committee is responsible for sorting votes, recommendations of the corresponding design project "graduation designing grade award" ranking.
(3)Finally, the guest judge has to recommend to base on the " graduation designing grade award " program ranking, and combine the results of the control panels with the review defense program, inviting the judges to fill in the ballot again (Universities judges must be avoided), and have a sorted selection. Organizing committee is responsible for sorting ballots statistics and forming the corresponding design themes of "graduation designing grade award" program selection results, and then considering in determining the winning grade.
3. The selection of the "Outstanding graduate Excellence Award"
The out-of-school judges and the school judges shall suggest the "Outstanding graduate Excellence Award" program Sort and vote, whose has the two-thirds number of the judges shall be
The praise rules of "Grade Award of the Graduation Design" and "Outstanding Graduate Excellence Award"
1．The CIID issue certificate to the winners of the "grade award of the graduation design", "Outstanding graduate Excellence Award"
2．The award certificates are designed and printed by the CIID and stamped with the seal of the CIID, which also has the signature of the judges of the committee.

The CIID
March 30, 2015

答辩规则（2015版）

一、毕业设计参加高校毕业设计答辩、展出准备要求

1. 参加高校每校到场指导教师不超过2名，到场学生总人数限6名。

2. 参加高校每校优选2个毕业设计答辩方案组，参加毕业设计答辩活动、"毕业设计等级奖"评选和成果展出；此外，最多可再报送2个自荐方案，参加"毕业设计优秀奖"评定和成果展出，但不参加毕业设计答辩和"毕业设计等级奖"评选。

3. 每个毕业设计答辩方案组提前准备毕业设计答辩陈述PPT等电子文档，于毕业设计答辩活动报到时提交活动组委会。

4. 每个毕业设计答辩方案和自荐方案需提前准备成果展板3张；展板幅面为A0加长：900 mm×1800mm，分辨率不小于100dpi，使用学会提供的统一模板编辑。展板电子版须于答辩活动前1周发送到答辩活动承办高校指定的工作邮箱；由承办高校负责汇总打印、布展等。

5. 参加高校按《CIID "室内设计6+1"校企联合毕业设计联合毕业设计书稿排版要求》编辑书稿，于毕业设计答辩活动现场提交活动组委会。

6. 毕业设计专题方向选报

"室内设计6+1"校企联合毕业设计的参加高校，结合本校毕业设计教学实际，分别按照当届《"室内设计6+1"校企联合毕业设计框架任务书》设置的毕业设计专题，对2个毕业设计答辩方案设计方向进行选报（可多选），并组织毕业生进行答辩。

二、毕业设计答辩与评奖规则

1. 答辩组评委由特邀评委和高校评委组成

（1）特邀评委由学会（包括活动地区地方专业委员会）、命题企业、支持企业等在内的5~7位专家担任；

（2）高校评委由参加高校各推选1位毕业生指导教师担任。

2. "毕业设计等级奖"评选

（1）"毕业设计等级奖"按当届《"室内设计6+1"校企联合毕业设计框架任务书》设置的毕业设计专题分别设置一、二、三等奖；一至三等奖的奖项设置比例一般为1:2:4；等级奖可空缺，空缺奖项数量在相邻等级间做调剂。

（2）首先，通过答辩，由特邀评委和高校评委共同评提出建议"毕业设计等级奖"方案排序。参加高校每个毕业设计答辩方案组按选报设计专题进行答辩。每组答辩陈述时间不超过15分钟，问答不超过15分钟。两类评委填写选票，进行排序评选（如，1为建议排序第一，2为建议排序第二，依次类推）。活动组委会负责排序选票统计，形成相应设计专题的建议"毕业设计等级奖"方案排序。

（3）最后，由特邀评委以建议"毕业设计等级奖"方案排序为基础，对照答辩方案成果展板进行审议，特邀评委再次填写选票（高校评委须回避），进行排序评选。活动组委会负责排序选票统计，形成相应设计专题的"毕业设计等级奖"方案评选排序结果，并审议确定获奖等级。

3. "毕业设计优秀奖"评选

特邀评委和高校评委共同对参加成果展示的高校自荐方案成果展板进行评议，对是否认定为"毕业设计优秀奖"方案进行投票；同意票数超过两类评委总人数2/3的高校自荐方案，确定获得优秀奖。

三、"毕业设计等级奖"、"优秀毕业优秀奖"表彰规则

1. 在活动颁奖典礼上，由学会分别向"毕业设计等级奖"、"优秀毕业优秀奖"获得者颁发证书。

2. 获奖证书由学会盖章，并有活动特邀评委和高校评委签名。

中国建筑学会室内设计分会
二〇一五年五月三十日

CIID"室内设计 6+1"2015（第三届）校企联合毕业设计
CIID "Interior Design 6+1" 2015(Scssion)University and Enterprise Joint in Graduation Design

调研踏勘
Reconnaissance Survey

调研踏勘 Reconnaissance Survey

CIID "室内设计 6+1" 2015（第三届）校企联合毕业设计
CIID "Interior Design 6+1" 2015(Third Scssion)University and Enterprise Joint in Graduation Design

同济大学
Tongji University

小组成员：张黎婷 Zhang Liting　马潇潇 Ma Xiaoxiao
　　　　　吴晓飞 Wu Xiaofei　　马曼·哈山 Maman·Hashan
　　　　　周　怡 Zhou Yi　　　　陈　杰 Chen Jie

文本调研
Document Investigation

首先，我们进行大量的关于工业遗产建筑改造的文本进行阅读和调研，提出核心问题及一些现有的问题，得出对工业历史建筑进行改建设计的两个基本特点：

第一是充分利用历史建筑原有的结构部分，根据要置入的新使用功能的需要把新的空间插入到原来的建筑中。

第二是保持历史建筑原有外观上的特征，对建筑立面的改变一般都遵循历史建筑立面原来的逻辑关系。

在这两个基本特点的影响下，建筑师常常寻找一种最合理、最简单甚至是最直接的方式把现代的建筑空间组合到历史建筑中去。

Firstly, we do plenty of read and investigation with documents of industrial heritage architecture reconstruction, and find key issues and raise current questions. Then, we come to a conclusion that there are two fundamental features of constructing industrial historical architecture:
Make full use of the original parts of the structure of historical architecture, and insert the new space into original architecture according to new functions.
Keep the original appearance characteristics of historical architecture. Change to building elevation are generally following the primary logical relations of the historical building elevation.
Architects usually search for the most reasonable, simplest and even the most direct way to combine the modern architecture space into the historical architecture by the effect of these two features.

"多维视角的规划和思考是解决历史街区可持续发展的必要条件，街区再生是物质空间、人文空间和社会空间的共同再生体。从街区整体性出发，将再生规划与这一地区的经济、社会、文化、民俗等各方面联系起来，避免僵化的功能分区，改变以往靠用地性质单一和过渡管制实现的视觉和谐和功能有效的思维模式，街区既要保持原有社会空间形态，也要适当置入或更换新的功能使新旧并存。"——摘自《弄堂工厂历史街区再生意义》

"The necessary condition to solve the sustainable development of the historic districts is multidimensional view of planning and thinking. And change the thought mode that comes from the visual harmony and function effectively leading by single land usage and excess control in the past. Districts should not only keep the original social space form, but also should be brought in new features appropriately to make the old and new coexist." --selected from "Significance of the land factory historical district regeneration"

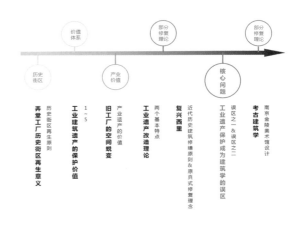

南京文脉解读
Nanjing Context Interpretation

通过百年南京记录片和档案史料的阅读，我们发现南京的历史十分跌宕起伏，它是六朝古都，同样也是民国政府的首要之地，在传统文化与现代文明之间的挣扎起伏，而且南京历史上的重大事件，导致多次人口剧变，这种历史发展是跌宕起伏的。

We found that Nanjing history was fluctuant violently by Nanjing in one century documentary and reading of historical data. Nanjing was the ancient capital of six dynasties of China, and was also the center place of the Government of the Republic of China, wandering between traditional culture and modern civilization. The important events in Nanjing history lead the population to change many times. The history development of Nanjing is full of ups and downs.

案例调研
Case Research

上海 M50 创意园区——本是一片被遗忘了的旧厂房和破损不堪的上海老式民房；这里的破旧与古老似乎已与发展中的城市格格不入。之后被上海几个知名艺术家租来作画廊，马上身价百倍，现大部分为艺术家和艺术民工。

M50 Creative Park in Shanghai was a forgotten place of old factories and broken residences, out of tune with the developing city. It was rented by several famous artists of Shanghai as studio, and is worth hundreds times now being crowded with artists and art workers.

基地区位分析
Base Location Analysis

交通便捷：园区靠近内环南线和动线交汇处，周边有地铁1号线、3号线、8号线及多辆公交线路。

Convenient transportation: The Park is near the intersection of inner ring south and circulation, with metro line 1, line 3, line 8, and several bus lines around.

旅游景点集聚：基地位于南京历史文化旅游景点集聚区，周边文化氛围浓厚，旅游文化区位特质明显。

Tourist attractions gathered: It is located in the area of historical and cultural tourist attractions in Nanjing surrounded by cultural atmosphere and the location characteristics of tourism culture are obvious.

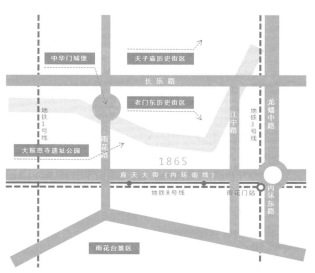

调研踏勘 Reconnaissance Survey

CIID "室内设计 6+1" 2015(第三届)校企联合毕业设计
CIID "Interior Design 6+1" 2015(Third Scssion)University and Enterprise Joint in Graduation Design

华南理工大学
South China University of Technology

小组成员：廖喆璇 Liao Zhexuan　郑宇晨 Zheng Yuchen
　　　　　周靖涵 Zhou Jinghan　马仓越 Ma Cangyue
　　　　　廖绮琳 Liao Qilin　　俞　快 Yu Kuai

背景分析
Background Analysis

中国　　　　江苏省　　　南京
CHINA　　　JIANGSU　　NANJING

南京，是江苏省会，地处中国东部地区，长江下游，濒江近海。南京有着6000多年文明史、近2600年建城史和近500年的建都史，是中国四大古都之一，有"六朝古都"、"十朝都会"之称。场地所在的秦淮区是南京市的中心城区。

Nanjing is the capital of Jiangsu Province. It is in the east of China, located in the lower reaches of Yangtze River, close to the river and sea. Nanjing is one of China's four ancient capitals with more than 6000 years civilization history, nearly 2600 nearly city history and nearly 500 years capital history. It is called "the ancient capital of the six dynasties", or "ten dynasties capital". The exploration field in the Qinhuai district is the central city of Nanjing.

– 南京市工业遗产分布 –　　　　　– 南京市创意产业园分布 –
-Nanjing Industrial Heritage Distribution-　-Nanjing Creative Industry Park Distribution-

南京作为民国时期的首府，在当时吸引了众多的企业家投入了大量的财富以及技术到工业发展中。其中保留至今的工业遗产随着南京的发展遍布市区内外，这些众多的工业遗产年代跨度大、遗存价值高、类型覆盖广、布局有章可循。1865晨光创意产业园作为晚清时期的工业厂房被安置在了明城墙以外。很多工业遗产都被顺理成章地改建为创业产业园，1865晨光创意产业园也不例外。

As the capital during the period of the Republic of China, Nanjing attracted numerous entrepreneurs at that time to invest a lot of wealth in the industrial development. The industrial heritages were preserved throughout the city with the development of Nanjing 1865 Chenguang Creative Industry Park was placed outside the Ming city wall as the industrial plant in the late Qing Dynasty. A lot of industrial heritages were naturally converted into industrial park including the 1865 Chenguang Park.

园区分析
District Analysis

园区位于南京市中心地区南侧（明城墙外的南侧），处于历史文化氛围十分浓郁的秦淮区，西邻报恩寺，东接江宁路口。新建地铁紧挨场地南端，101路、16路、202路等十几条公交线路途经场地，地理位置优越，交通也十分便利。

The Park is located in the south of central area of Nanjing (outside the Ming wall south), belonging to Qinhuai district where the historical and cultural atmosphere is very strong. Hoonji is in its west and Jiangning road in its east. The new subway is next to the south of the Park. And there are bus 101, 16 and 22, dozens of bus lines through around the Park. So its position is good, so as the transportation.

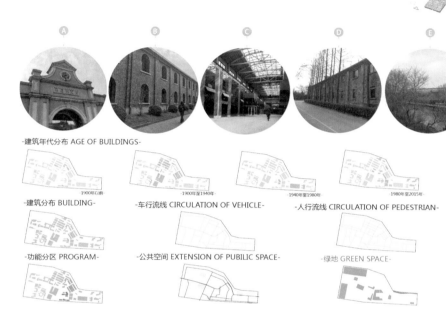

- 建筑年代分布 AGE OF BUILDINGS -
- 建筑分布 BUILDING -
- 车行流线 CIRCULATION OF VEHICLE -
- 人行流线 CIRCULATION OF PEDESTRIAN -
- 功能分区 PROGRAM -
- 公共空间 EXTENSION OF PUBLIC SPACE -
- 绿地 GREEN SPACE -

产业园所在地为李鸿章于1865年兴建的金陵制造局旧址，园区有9幢清代建筑、19幢民国建筑，如同一座近代中国工业博物馆，记录着中国民族工业发展的历史轨迹。

The place where the Park is located is the former site of Quanling manufacture bureau built by Li Hongzhang in 1865. There are 9 buildings of Qing Dynasty and 19 buildings of the Republic of China in the Park. It is like a museum of modern China's industry, recording the historical track of the development of the national industry.

建筑结构分析
Original Analysis

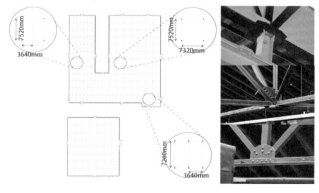

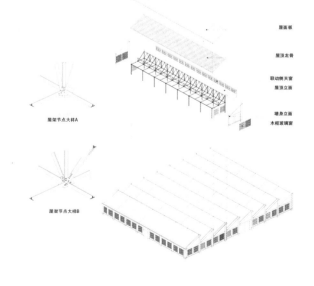

A1及A2始建于民国时期，为钢混结构建筑，单元钢架结构不断重复，外围为砖石结构的墙体。A1建筑钢架结构为均匀排布的7520毫米×7320毫米的钢架结构，A2建筑钢架结构存在三种规格的柱跨。

A1 and A2 were reinforced concrete structure built in the period of the republic of China. They use the steel frame structure unit repeatedly, with the external wall of masonry. A1 is 7520mm× — 7320mm steel frame structure evenly arranged. A2 is steel frame structure with the column span of three kinds of specification.

CIID "室内设计 6+1" 2015（第三届）校企联合毕业设计
CIID "Interior Design 6+1" 2015(Third Scssion)University and Enterprise Joint in Graduation Design

哈尔滨工业大学
Harbin Institute of Technology

小组成员：金　喆 Jin Zhe　　　何佳佳 He Jiajia
　　　　　伏　祥 Fu Xiang　　　张泽宇 Zhang Zeyu
　　　　　乔　红 Qiao Hong　　 韩思宇 Han Siyu
　　　　　贾思修 Jia Sixiu　　 王　岩 Wang Yan
　　　　　王子轩 Wang Zixuan

基地介绍
Introduction of the Base

1865创意产业园区是南京规模较大、较为成功的工业建筑改造群。该园区最早是清朝晚期洋务运动时李鸿章所创建的金陵制造局。该地区自创建以来至今，已经连续运作了近150年，留下了各个时期发展的痕迹，是中国近代机器制造业发展历程的一个浓缩。

In Nanjing, 1865 Creative Industry Park is a successful rebuilding architecture group which has a big scale. The park's history can trace back to the Westernization Movement of late Qing dynasty. It was Jin Ling Manufacturing Bureau built by Li Hongzhang. The park has been run nearly 150 years and left development signs of all stages. It is a potted history of the development of modern China's machine-building industry.

园区占地面积21万平方米，总建筑面积11万平方米，产业总体定位为文化艺术和创意设计，重点引进文化传媒、艺术创作、创意设计、科技研发、总部经济企业、高端人才等。

The park covers 210,000 m^2, with a total floorage of 110,000 m^2. The overall positions of the industry are cultural arts and creative design. It puts priority on introducing cultural media, art creation, creative design, technology research and development, headquarters economy enterprises and high-end talents, etc.

园区位于南京市中心地区南侧，距夫子庙仅1.2公里。北临秦淮河，与中华门城堡隔河相望，园区周边交通便利，南京井字形城市快速通道之一的纬七路从东西穿越。距禄口机场仅38公里，距地铁中华门站约500米；同时十几条公交线路分别与大明路、中山南路、城东干道、城西干道等多条城市主干道及机场高速公路相连接，交通便利。

The park located in the south of the central area of Nanjing, 1.2 km from Confucius Temple, with Qinhuai River to the north and the Zhonghua Gate across the river. The transportation around the park is very convenient. Wei 7 Road, one of the Nanjing's well-shaped expressways, crosses the park from east to west. It locates 38 km away from Nanjing Lukou International Airport and 500 meters away from Zhonghua Gate subway station. More than 10 bus lines connect with many city main roads and airport expressway, such as Daming Road, Zhongshan South Road, and main roads of the east and west of the city.

园区现状
The Present Situation of the Park

A1、A2 馆介绍
Introduction of the Exhibition Hall A1 and A2

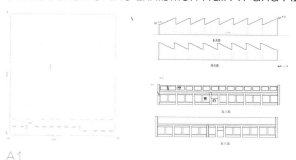

A1

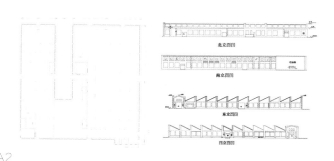

A2

始建于民国时期（1934年），一层钢混结构，建筑面积3309.6平方米，原为软管公司工房。房屋长×宽×高为62.1米×53.1米×5.77（9.17尖顶）米。

A1 was built in the Republican period (1934). The first floor is steel-concrete structure with the floorage of 3309.6 m². It used to be the processing room of a hose company. It's L*W*H are 62.1*53.1*5.77 meters (the fastigium is 9.17).

始建于民国时期（1934年），一层钢混结构，建筑面积8236.3平方米，原为12分厂机加工房、后为软管加工工段。房屋长×宽×高为92.8米×93.3米×5.77（9.17尖顶）米。

A2 was built in the Republican period (1934). The first floor is steel-concrete structure with the floorage of 8236.3 m². It used to be the machining room of the branch plant 12, and then the hose processing section. It's L*W*H are 92.8*93.3*5.77 meters (the fastigium is 9.17).

调研踏勘 Reconnaissance Survey

CIID "室内设计 6+1" 2015（第三届）校企联合毕业设计
CIID "Interior Design 6+1" 2015(Third Scssion)University and Enterprise Joint in Graduation Design

西安建筑科技大学
Xi'an University of Architecture and Technology

小组成员：冯胤云 Feng Yinyun　　何其畅 He Qichang
　　　　　韩杰林 Han Jielin　　　黄　超 Huang Chao
　　　　　胡铭中 Hu Mingzhong　　刘璧凝 Liu Bining
　　　　　孙宗藜 Sun Zongli　　　李肖路 Li Xiaolu
　　　　　潘　岩 Pan Yan

金陵风吟
Wind Sing of Jinling

项目位于南京秦淮区正学路1号；
项目占地面积21万平方米，总建筑面积9万平方米；
产业园所在地为李鸿章于1865年兴建的金陵制造局旧址；
园区致力于建造成国内知名的融科技、文化、旅游、商业等为一体的综合性生活地标和创意产业中心

The project is located in No. 1, Zhengxue Road, Qinhuai District, Nan
The project covers 210,000 ㎡, with a total floorage of 90,000 ㎡.
The place where the park is was Jin Ling Manufacturing Bureau buil
Li Hongzhang in 1865.
The park tries to make an outstanding domestic integrated life landm
and creative industry center which merged with technology, cult
tourism and commerce together.

建面：4799平方米　建筑：4栋民国1栋建国．功能：从事艺术
的专业人士建立个人工作室。
Floorage: 4,799 ㎡ Building: 4 in Republican period and 1 after
establishment of PRC.
Function: establishing personal studio for the professionals of arts.

建面：13974平方米　建筑：4栋，其中绝对调高14米的大空间3
功能：创意及艺术品展示，高科技及创意公司办公。
Floorage: 13,974 ㎡ Building: 4 buildings, three of which absolute he
adjustment are large spaces of 14 meters.
Function: creativity and artworks exhibition and high-tech and crea
business running.
科技创意研发区
Technological creative D & R

项目借鉴点 The reference of the project

✓ 优点：
园区开发规模大，保留原生态景观和古朴建筑群，
人文历史气息浓厚；
整个园区功能分区鲜明，每个企业入住对应的创意
区，形成各自的特色区域。

🚫 缺点：
园区占地面积较大，园区环境维护量较大，死角较多；
周边交通复杂，不便利；
沿河绿地利用率不高，可设置公共休闲区；
内部导识系统不全；
在创意设计上，园区还是保留原有景观为主，景观改造较少

Advantages

With a large scale of development, the park maintains a lot of original ecological landscape and ancient builds group, which owns a rich humane and historical atmosphere.
The whole park has a clear functional division. Each company will be arranged into the related creative part and form respective characteristic zone.

Disadvantages

The park covers too much land size to keep good maintenance and there are too many blind sites.
Its surrounding transportation is very complicated and not convenient.
The greenbelt along the river is not well used. It could be set as public leisure zone.
The internal guiding system is uncompleted.
On creative designing, the park still maintains the original landscape and is lack of rebuilding the landscape.

工业遗址的起源探索和文化创意产业园区的定义
The research of the development of the original industry location and the definition of the creative industry park

ORIGIN

20 世纪 60 年代英国工业考古研究的兴起。

The rise of industry archaeological research in the 1960s in Britain.

工业遗产的更新内容涵盖工业区整体更新、单体建筑更新、设施更新及环境更新 四个主要方面。工业区整体更新的方式主要有三种类型。

The rebuilding industrial heritage includes four main aspects: the whole industrial estate, single building, equipments and the surroundings. There are three main types of rebuilding ways of the whole industrial estate.

NAN JING

初期设计成几大功能分区，并不专营某项，实际操作后以商业、办公为主，餐饮、住宿、艺术为辅混杂入驻，呈现出多样化的功能组织，并且环境的附加价值对招商起到了显著的作用。

In early designing, we will design several functional zones which not only operate one project. In practical operation stage, it bases on commerce and offices while introducing catering, accommodation and arts subsidiary. It presents diversified functional merchants. The added value of the surroundings plays a significant role in recruiting investments.

before → now

2003 年国际工业遗产保护协会颁布了《下塔吉尔宪章》

The Nizhny Tagil Charter for the Industrial Heritage published by TICCIH in 2003.

我国的工业遗产保护起步较晚。政府主导，向城市高端化、专业化并拢。虽然企业搬迁对城市整合资源、调整结构、转型升级有积极作用，但城市文脉不可避免地出现了裂痕。

The starting of our country's industrial heritage protection is very late. Leading by the government, it goes to city high-end and professionalized. Although the companies' moving have positive effects on helping the city to integrate resources, restructure the structure, transform and upgrade, the city's history carrier cannot avoid growing a rift.

CHINA

国际工业遗址改造进程
The process of international industrial heritage rebuilding

小结

旧建筑的改造原因归纳起来有以下四点：经济因素、社会文化因素、情感因素、环境因素。这四点是非常重要的。其次，通过现代主义、后现代主义、解构主义、极少主义、有机形态等艺术手法都在旧建筑上有过尝试。随着法国卢浮宫的改扩建，德国国会大厦的改建成功，标志着现代主义手法在旧建筑利用上的成熟和主导地位。

我们在资料中总结出了几种常见的旧建筑改造方式：① 变换功能；② 化整为零；③ 结构改造；④ 扩建改造。

该调研报告基于实地调查的基础上，对旧工业建筑的改造方面总结出了要根据建筑环境等不同的因素进行相应准确的改造策略，从一个方面展现了不同旧建筑改造的需求，这样可以避免盲目跟风不求变化的改造方案，以此契机充分发挥不同旧工业遗产内在的价值与作用，进一步延续工业建筑遗产的精神与使用寿命。

Summary

There are four important reasons to rebuild ancient buildings: economy, social culture, emotion and environment. Secondly, the art practices like Modernism, Post-modernism, Deconstructionism, Minimalism and Biomorphism were used on ancient buildings. With the expansions of the Louvre, France and the success of the Reichstag rebuilding, it marks the ancient buildings of the Modernism's mature and dominated usage.

We conclude several common methods of rebuilding ancient buildings.
1) Changing function   2) Breaking up the whole into parts
3) Structure reconstruction   4) Transformation and expansion

This report is based on field survey, proposes remolding strategies on ancient buildings. It suggests that we should make relevant strategies and reflects different needs of ancient buildings from one aspect according to the different factors. So that we can avoid making our plan follow blindly. Taking this chance, the inner value of the industrial heritage will be fully worked. Furthermore, it will keep the spirit of industrial heritage and extended its serving life.

调研踏勘 Reconnaissance Survey

CIID "室内设计6+1" 2015（第三届）校企联合毕业设计
CIID "Interior Design 6+1" 2015(Third Scssion)University and Enterprise Joint in Graduation Design

北京建筑大学
Beijing University of Civil Engineering and Architecture

小组成员：郗嘉琪 Xi Jiaqi　　　姜　帅 Jiang Shuai
　　　　　刘凯南 Liu Kainan　　顾文博 Gu Wenbo
　　　　　付晨辉 Fu Chenhui　　王曌伟 Wang Zhaowei
　　　　　余祺盈 Yu Qiying

北京创意产业园调研
The Survey of Beijing Creative Industry Park

首钢艺术区
Shougang Group Art Zone

从北京市区开车向西，一出西五环，三四十米高的褐色炼铁高炉和烟囱尽收眼底，这里就是石景山老厂区，现在是3A级旅游景区。

Drive to the west from the downtown of Beijing, you can have a great view of some brown iron refining blast furnaces and chimneys which are 30-40 meters high as you get out west of fifth ring. Here is the former site in Shijingshan, and now it's a class 3A tourist attraction.

中间建筑
Middle Construction

中间建筑位于北京市海淀区，是一家民营美术公益机构。

Middle Construction is located in Haidian District of Beijing. It is a privately operated fine arts organization which faces to the public.

新华1949
Xinhua 1949

新华1949位于北京西城区黄金地段，是北京老城区一道风景线。

Xinhua 1949 is located in the prime location of Xicheng District, Beijing. It's a great view of the old Beijing city.

798
798

798艺术区，又称大山子艺术区。现已成为了北京都市文化的新地标。

798 Art Zone, also known as Dashanzi Art zone, has become the new landmark of Beijing urban culture.

二十二院街艺术区
The Twenty-two street Art Zone

北京二十二院街艺术区是北京最具人文气质的艺术型街区。

The Beijing Twenty-two Street Art Zone is an art street which has a great sense of humanity and culture.

南京 1865 方面调研
Survey of 1865 in Nanjing

1865创意产业园南侧有城市主要干道应天大道，高铁车站距该区域7公里，与建设和规划中的南京地铁3号线与8号线在东侧交汇。
To the South of 1865 is the main city street– Yingtian Street. The high speed train is 7km away from it and meets the subway line 3 and line 8 at the east side which is in construction and planning.

晨光1865所占区域
The area Chenguang 1865 occupied

● 地铁3号线 Subway line 3　　一般道路 Common road
● 地铁8号线 Subway line 8　　快速路 Express way

晨光1865空间环境
Space environment of Chenguang 1865

1865创意产业园位于南京中心地区南侧的秦淮区，距离夫子庙1.2公里，北临秦淮河，与中华门城堡隔河相望。

The 1865 Creative Industry Park is located in Qinhui district which is to the south of the center of Nanjing. It is 1.2km away from the Temple of Confucius. Qinhuai River is to its north and across this river is the Zhonghuamen Castle.

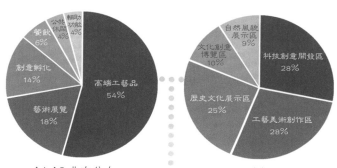

A1 A2 业态分布
A1 A2 Distribution of industry

园区功能分布
Function distribution of the park

■ 科技创意开发区
■ 工艺美术创作区
■ 历史文化展示区
■ 文化创意博览区
■ 自然风貌展示区

新建建筑　历史建筑　风貌建筑
New construction　Historical construction　Feature construction

人流分析
Analysis of people stream

动静分析
Analysis of movement and quiescence

● 静　　● 动

调研踏勘 Reconnaissance Survey

CIID "室内设计 6+1" 2015（第三届）校企联合毕业设计
CIID "Interior Design 6+1" 2015(Third Scssion)University and Enterprise Joint in Graduation Design

南京艺术学院
Nanjing University of the Arts

小组成员：陆　路 Lu Lu　　黄文进 Huang Wenjin
　　　　　刘曼羽 Liu Manyu　蔡文镇 Cai Wenzhen
　　　　　陈飞宇 Chen Feiyu　许　超 Xu Chao
　　　　　杨广智 Yang Guangzhi　刘佳俊 Liu Jiajun
　　　　　杨　鸣 Yang Ming

周边环境分析
Surrounding Environment Analysis

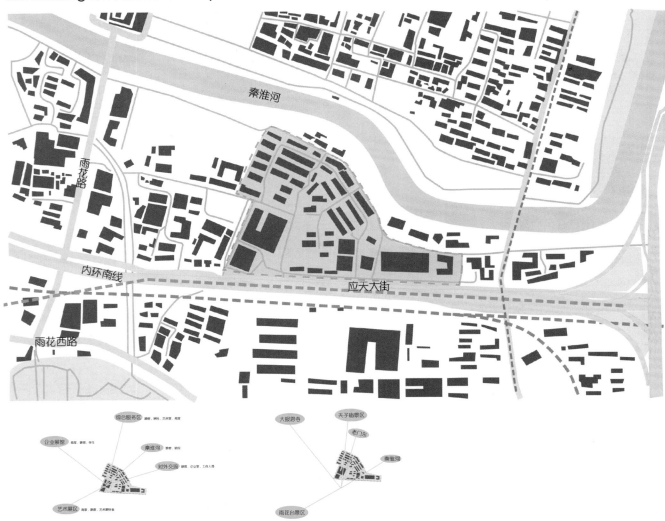

实地勘察分析
Field Survey Analysis

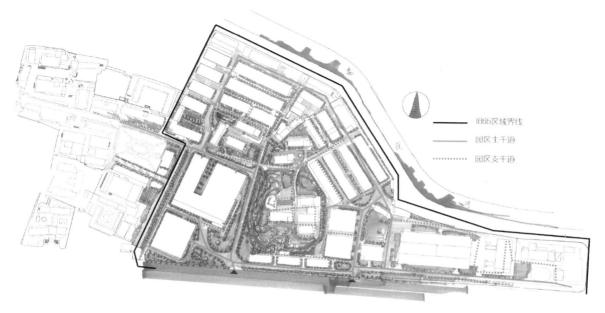

园区内部交通分析图
Traffic analysis chart of the internal park

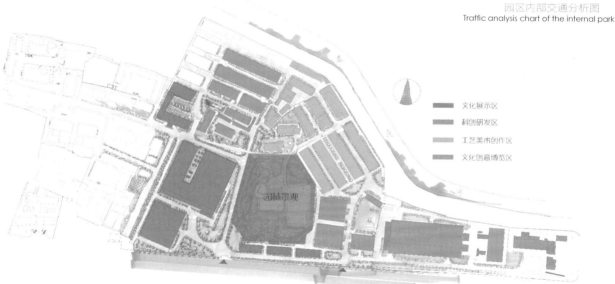

园区内部功能分析图
Functional analysis chart of the internal park

优势因素
Advantages
1. 园区位于南京市中心区南侧，交通便利，周边地铁、城市干道与高速路等形成便捷的交通运输网络。
2. 周边地区蕴含深厚的文化历史底蕴，自然景观丰富，吸引人群汇集。
3. 园区内保留了很多不同历史时期的建筑，具有较高的保护和利用价值。

1. The park is located in the south of the centre of Nanjing. The convenient transportation network connects with subway, city main roads and expressways, etc.
2. This zone is enriched with culture and history. The abundance of natural landscape attracts the audience gathering.
3. The park maintains many buildings of different periods of history, which has high value of protection and usage.

项目劣势
Disadvantages
1. 内部各个区域间联系不紧密，导视系统不全。
2. 历史变迁中工业文化未得到有效的传播。

1. Each internal part is not closely connected to each other, and the guiding system is uncompleted.
2. The historical industry culture is not well spread.

机遇与挑战
Challenges and Opportunities
1. 创意经济市场需求旺盛，具有积极的产业政策支持。
2. 根据现状条件，如何对园区进行正确定位将影响园区的发展前景。
3. 如何在保护性开发利用老建筑的基础上进行创新改造，是我们面临的最大的挑战。

1. High demand of creative economy market has positive industrial supporting policies.
2. According to the present situation, how to correct positioning the park will influence its development.
3. How to create and rebuild the ancient builds on the base of protective development is the greatest challenge we faced.

调研踏勘 Reconnaissance Survey

CIID "室内设计 6+1" 2015（第三届）校企联合毕业设计
CIID "Interior Design 6+1" 2015(Third Scssion)University and Enterprise Joint in Graduation Design

浙江工业大学
Zhejiang University of Technology

小组成员：张泽浩 Zhang Zehao　　夏　欣 Xia Xin
　　　　　马凯杰 Ma Kaijie　　　　孙　莹 Sun Ying

历史文脉分析
History Context Analysis

1865　　　　1877　　1937　　　　1934年　1962年　　　　2007年

1865—2007年（金陵兵工厂改造前）　　　　2007年开始改造1865创意产业园（改造后）

区位分析 Location Analysis

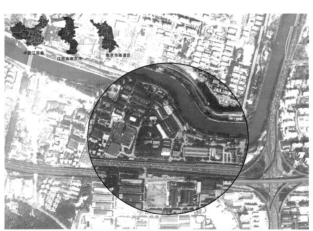

园区定位 Park Positioning

■ 科技创意研发区
■ 山顶酒店商务区
■ 时尚生活休闲区
■ 工艺美术创作区
■ 科技创新博览区

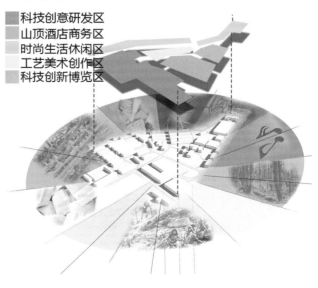

园区现状分析 Park Status Analysis

—— 人行道 Sidewalk

----规划区域 Buildings Distribution
■ 建筑分布 in Planning Region

绿化层 Greening Layer

➡ 大门入口 Gate Entrance
—— 车行道 Driveway

基地周边 Base Surrounding

基地人流量 Base passenger flow
■ > ■

客群分析 Customers Analysis

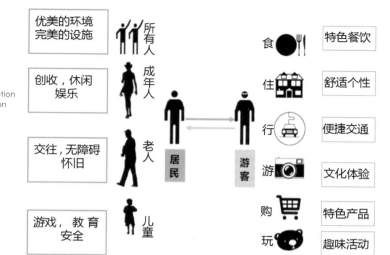

交通问题 Transportation Problems

狭窄、闭塞、密集的道路与大量车流、人流间的矛盾。

Contradiction among the narrow and blocked roads and the large influx of vehicles and people.

生态问题 Ecological Problems

建筑的高密度，空间系统的凌乱，不完整园区缺乏大量公共空间。

Buildings of high density, messy space system, the incomplete Park lacks of a large number of public spaces.

空间结构 Spatial Structure

园区生态绿地分布不均匀，绿地面积不足，绿地之间缺少联系。

The Park greens are ill-distributed, area-lacking, and independent from one another and lacking of connections.

周边渗透 Surrounding Penetration

厂区各个部分之间相对独立，缺少联系。

All parts of the factories are relatively independent, lacking of connections.

CIID "室内设计 6+1" 2015（第三届）校企联合毕业设计
CIID "Interior Design 6+1" 2015(Third Scssion)University and Enterprise Joint in Graduation Design

过程方案
Process Plan

过程方案 Process Plan

CIID "室内设计 6+1" 2015(第三届)校企联合毕业设计
CIID "Interior Design 6+1" 2015(Third Scssion)University and Enterprise Joint in Graduation Design

同济大学
Tongji University

小组成员：张黎婷 Zhang Liting　马潇潇 Ma Xiaoxiao
　　　　　吴晓飞 Wu Xiaofei　　马曼·哈山 Maman·Hashan
　　　　　周　怡 Zhou Yi　　　　陈　杰 Chen Jie

金陵印　Jinling Seal

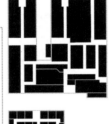

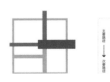

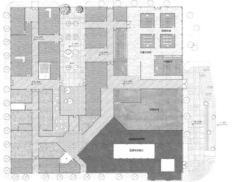
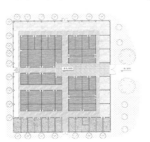

平面图

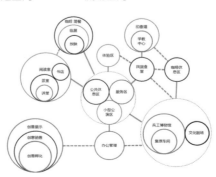

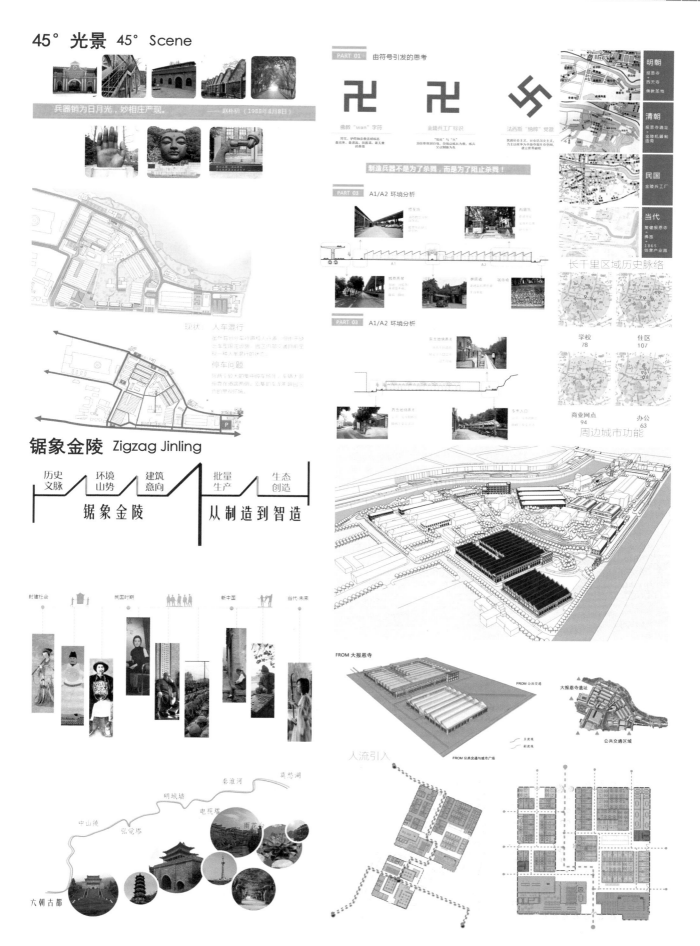

CIID "室内设计6+1" 2015（第三届）校企联合毕业设计
CIID "Interior Design 6+1" 2015(Third Session)University and Enterprise Joint in Graduation Design

华南理工大学
South China University of Technology

小组成员：廖喆璇 Liao Zhexuan　郑宇晨 Zheng Yuchen
　　　　　周靖涵 Zhou Jinghan　马仓越 Ma Cangyue
　　　　　廖绮琳 Liao Qilin　　俞　快 Yu Kuai

概念生成
Concept Generation

我们将设计定位为一个对公众开放的文化体验中心，同时我们开始思考如何让这个旧的厂房产生新的活力。我们希望在这个厂房内加入新的体块，这些体块代表了新的功能、新的技术革命，于是在同一空间中，延续了两次革命的对话。

We position the design as a cultural experience center which is open to the public. Meanwhile, we begin to think how to make the old plant lively. We hope to insert new part into the plant, which represent the new functions and new technical revolution. So that in one space, two revolutions continue.

旧空间　　　兵工技术革命

VS.　←------→　VS.

新体块　　　新技术革命

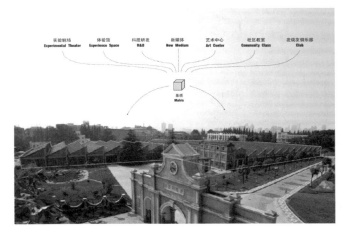

Process Plan 过程方案

流线分析
Streamline Analysis

图底关系
Graph Base Relation

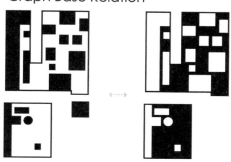

设计将 A1、A2 两个厂房串联作为主要的空间串联，从旧厂房以及新加入的体块的图底关系我们可以限定出介于两者之间的一种新的空间，这些空间作为流动空间，可以被赋予不同的功能，根据不同的活动和需求进行不同的布置。

The two plants, A1 and A2 are connected in series as the main space connection. We can define a new space from the figure and ground between the old plants and the new volumes. As flowing spaces, these spaces can be given different functions and can be arranged in various ways according to different activities and needs.

剖面分析
Profile Analysis

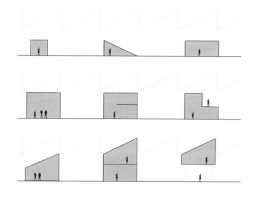

轴测分析及效果
Axonometry Analysis & Effects

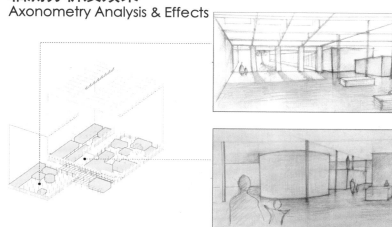

活动分析
Activity Analysis

效果示意
Effect Sketch

前期调研
Early Research

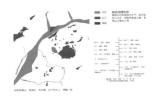

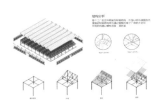
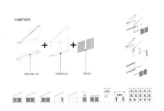

南京山川形势极为大气，前可挹长江之水，背则倚紫金之峰，兼得山之雄水之秀。无论是历史还是现在，南京都受到江南江北大部分地区的映射，在经济文化各个方面形成了一种独特的交融。

The geographical situation in Nanjing is extremely magnificent. There is the Yangtze River in front of the city and the Purple Mountain back of it, Nanjing enjoy itself natural beautiful. Whatever the past or the modern Nanjing, it has been mapped by most regions of the south and north of the river, forming a characteristic fusion in various aspects of economy and culture.

一个有趣的发现 An Interesting Discovery

以"卍"字符作为金陵兵工厂的标志这个字梵文读"室利踞蹉洛刹那"，意思是"吉祥海云相"，也就是呈现在大海云天之间的吉祥象征。它被画在佛祖如来的胸部，被佛教徒认为是"瑞相"，能涌出宝光，"其光晃昱，有千百色"。

The sign 卍, swastikais, used as the symbol of Jinling Arsenal, which is the auspicious symbol between the sea and the sky. It is one of the auspicious signs recognized as being on the chest of Buddha. It is considered as "Auspiciousness" by the Buddhists, pouring out precious light, "The light shakes, having thousands of colors."

一个有趣的轮回

金陵机器制造局
1869年，金陵制造局已能制造多种口径前膛炮、炮弹以及炮车和枪弹。在个张正中发挥了重要的作用。

西天寺
1865年，李鸿章选定在南京雨花台的东北侧，明朝西天寺旧址上建立金陵机器制造局

晨光集团
1980年3月改称南京晨光机器厂；1996年6月，作为全国一百家现代企业制度试点单位之一，正式组建了晨光集团，并更名为南京晨光集团有限责任公司。

佛　静谧　安详　硝烟　战火　释然　平静　沉静　战争
兵器　肃穆　从容　生活　怜悯　伤亡　失去　和平

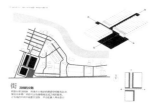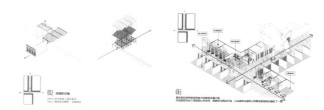

体量庞大的厂房建筑在场地上显得突兀，且中间区域的可达性差，活跃度低，将部分屋顶拆除，将原本分离的巷道空间联系起来。看似相交的两条街将庞大的建筑体量分离，但因为这个活跃的公共空间，实际又将建筑与周边环境以及建筑与建筑之间更加紧密地交融在一起，增加了大体量厂房的可达性，增强了各区域之间的联系。为使场地的中间更加活跃，在沿街置入商业部分。

The large volume plant buildings appear abrupt on the ground, and the middle area has poor accessibility and less lively. Dismantle part of the roofs and combine the original separated roadway space. The two streets seemed to be intersecting but separate the large volume buildings, but actually because of the active public space, they combine the buildings and the environment around, and between buildings as well more closely to increase the accessibility of the large volume buildings and the connection of the regions, making the middle area more active, and insert commercial elements along the streets.

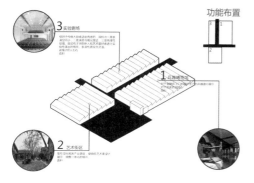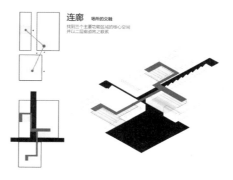

三个主要功能区域的核心空间以二层廊道作为联系，各庭院中以民营工厂时期所生产的佛像作为主题。宁静的天光、禅意的空间，与兵器的冰冷形成对比，带给人们历史在此空间交融的感受。

Find the core space of the three main function regions and connect them by the corridor of the second floor. Each courtyard takes the figure of Buddha produced during the private factories as the theme. The quiet sky and the space with artistic conception of Zen contrast with the cold of weapons, bringing people dramatic feeling. History becomes integrated in this kind of space.

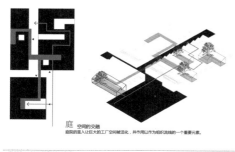

过程方案 Process Plan

CIID "室内设计 6+1" 2015(第三届)校企联合毕业设计
CIID "Interior Design 6+1" 2015(Third Scssion)University and Enterprise Joint in Graduation Design

哈尔滨工业大学
Harbin Institute of Technology

小组成员：金　喆 Jin Zhe　　　何佳佳 He Jiajia
　　　　　伏　祥 Fu Xiang　　　张泽宇 Zhang Zeyu
　　　　　乔　红 Qiao Hong　　 韩思宇 Han Siyu
　　　　　贾思修 Jia Sixiu　　　王　岩 Wang Yan
　　　　　王子轩 Wang Zixuan

理念透析
Concept Analysis

工业博物馆的特点：
1 馆内的藏品与工业有关
2 保护工业历史遗迹和弘扬工业历史文化

↓ 定位

回顾工业历史，展望科技未来 → 理念

满足博物馆基本展示功能（前提）
加强教育与服务功能（核心）
科技手段提高兴趣（方法）
成为中国军工业发展的窗口（目标）

↓ 对象

青少年、家庭团体、游客

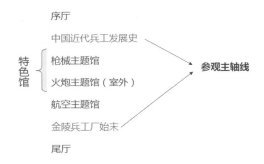

将其转变为一个对公众开放的综合性博物馆——近代兵工博物馆。它的设立，除了可以更好地保护遗产，还可以丰富晨光创意产业园的人文内涵，应该把它作为打造人文秦淮品牌的具体项目来操作。

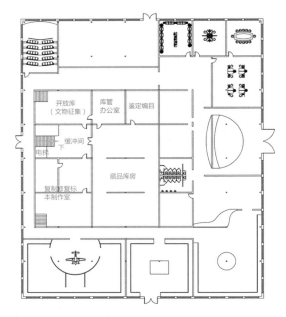

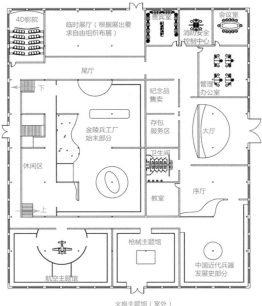

概念构想
Concepts and Conceptions

分析——问题
Analyze -- Problems

1. 工业遗址 + 商业 ———— 沉重而繁闹的状态长期带来的感受：严肃、压抑。
1. Industrial site + Business —— Its long-term heavy and bustling history making people feel: seriousness, suppression.

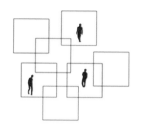

2. 部分人群对于区域会有选择性困难。
2. Some people have Selective difficulties on areas.

We will change it into a comprehensive museum, Modern Ordnance Museum, and open it to the public. Its establishment will not only preserve the heritage, but also enrich the humanistic connotations of Chenguang Creative Industry Park. It should be operated as a specific project of humanistic Qinhuai brand.

分析——解决
Analyze -- solutions

———— 刚与柔的结合
The combination of rigid and flexible

1. 引入绿色——使历史元素与现代理念的结合，形成历史向自然的过渡。
1. Introducing the green idea ——It combines historical elements with modern ideas, forming a transition from history to nature.

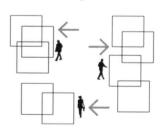

2. 对整体功能有明确的分类 —— 模式统一，形成连锁反应。
2. Clear classification of the whole function——unified pattern and forming chain reaction.

方案生成
Generating Schemes

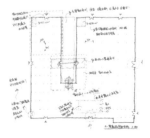 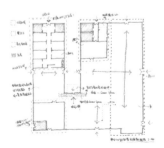 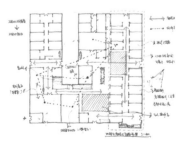 

CIID "室内设计 6+1" 2015(第三届)校企联合毕业设计
CIID "Interior Design 6+1" 2015(Third Scssion)University and Enterprise Joint in Graduation Design

西安建筑科技大学
Xi'an University of Architecture and Technology

小组成员：冯胤云 Feng Yinyun 何其畅 He Qichang
韩杰林 Han Jielin 黄 超 Huang Chao
胡铭中 Hu Mingzhong 刘璧凝 Liu Bining
孙宗藜 Sun Zongli 李肖路 Li Xiaolu
潘 岩 Pan Yan

区位分析
Area Analysis

园区位于南京市中心地区南侧，处于历史文化氛围十分浓郁的秦淮区，距夫子庙仅1.2公里，北临秦淮河，与中华门城堡隔河相望，距禄口机场仅38公里，离中华门站500米，规划中的地铁3号线和8号线在园区东部交汇，与拟建的亚洲最大火车站南京南站仅2.4公里。待园区所有的业态设计完毕会吸引大量的人前来参观、交流。瞻园、夫子庙、朱雀桥、中华门、大报恩寺、雨花台、古城墙等都在园区附近，地理环境优越，可以打造成综合性的经济产业带。

The park is located in the south of the center of Nanjing, where enriched in historical culture. It's only 1.2 km from Confucius Temple, with Qinhuai River to the north and the Zhonghua Gate across the river. It from Nanjing Lukou International Airport 38 km and the Zhonghua Gate subway station 500 meters. The planning subway line 3 and 8 will contact at the east of the park, which only 2.4 km from the proposed Asia's largest railway station, Nanjing South Station. After all the format designs are completed, the park will attract a lot of people to visit. Zhan Yuan, Confucius Temple, Suzaku Bridge, Zhonghua Gate, Da Baoen Temple, Terrace of Raining Flowers and the ancient city wall are all around the park. With the advanced environment, it could be created an integrated economy industrial belt.

降雨分析
Rainfall Analysis

通风与日照
Ventilation and Sunshine

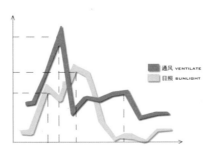

能耗分析
Energy Consumption Analysis

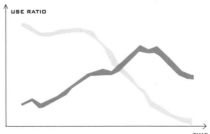

关系分析
Relationship Analysis

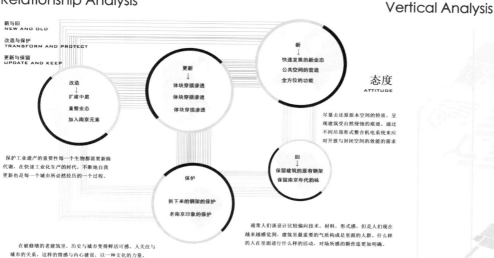

原有场地的"设计"反映过去人的工作和生活，以及当时的审美和价值取向，或许与现代人的需求有一定的距离。但经过改建，通过增与减的设计，在原有"设计"基础上产生新的形式，使能戏剧化地讲述场地的故事，揭示场所的精神，又能人性化地满足现代人的需求和欲望。保护并合理改造旧厂区中的工业遗产，是表达对历史的敬意，也是表达对现实的关怀。

The original design reflects people's work and life and their taste and values at that time. Maybe there is a distance from modern people's needs, but after reconstruction and the design of adding or cutting, it will create new form on the base of the original design. It will tell the park's story dramatically, show its spirit poetically and meet modern people's needs and desire with humanity. Protect and reconstruct industrial heritage reasonable is a way to salute history and show respects to reality.

竖向分析
Vertical Analysis

园区周边业态分析
Business Analysis around the Park

南京晨光医院	中旅宾馆	尼古拉会所
丁村小区	雨花路药店	随心圆酒家
晨光技工学校	先声再康药店	正大拍卖
荣鼎汇	玉光宾馆	双桥门
天喜餐饭会所	金陵大报恩寺塔	雨花情餐厅
南京晨光老年大学	友善旅社	南京秦淮区卫生监督所
金陵王谢	晨光宾馆	春味牌有机茶
妙奇汽车	1865锦绣秦淮	皇州茗茶
公正村	美好晨光生活超市	天天沙县小吃
梅花村	美好晨光中式快餐	南京雨花白癜风研究院
晨光巷小区	晨光公寓	

过程方案 Process Plan

经调研，园区周边5公里内业态包括药店2家、宾馆5家、小区6处、餐饮12家、古迹6处、商场4家、学校3所、医院4家、个体经营（除餐饮）13家。

We surveyed that there are 2 drug stores, 5 hotels, 6 communities, 12 restaurants, 6 historic sites 4 malls, 3 schools, 4 hospitals and 13 individual operators (exclude restaurant) within 5 km around the park.

提出问题 Raise Questions

解决办法 Solutions

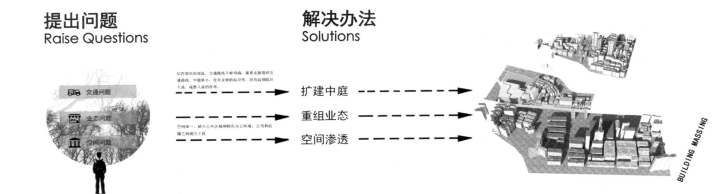

从空间角度出发，联系人与城市之间的关系，提取出具有典型胡同肌理的南京城市肌理。以盒子体块为元素穿插、扭转，创造出不同功能的空间，增加室内的公共空间和室内绿地，营造良好的环境。探寻设计和建造的自然性，通过尽可能少的元素传达旧建筑本身带给场所的静谧感和一种简朴甚至清苦的美学。

From the view of space, contact the relationship of human and city, we take typical Hutong of Nanjing urban tissue, using boxes as the element to create different functional spaces. We also create a graceful environment by adding indoor public and green space. We try to find the nature of design and construction, making as few changes as possible to express quiet and a simple or even austere aesthetics coming from the ancient builds.

概念生成 Concept Generation

将同样典型的北京城市胡同肌理与南京城市胡同肌理相对比，发现南京的海派建筑特点在于在每家每户出门前的水渠。提取水渠的元素，贯穿到建筑的空间划分和交通流线上，既更好地使空间进行了自然渗透，也引导了人流的活动方向。水流从上依次跌落下来，形成了灵动的水帘，给人们带来了全新的空间感受。

We contrast typical urban Hutong tissue between Beijing and Nanjing and found that there are ditches in front of every household in the style of Shanghai Building. We take the element of ditch to run through the partition of the building's space and the flow line of the transportation. It could not only infiltrate the space naturally, but also guide the flow of visitors. The water falls down and forms a living water curtain, giving the visitors an entirely new feeling of the space.

Process Plan 过程方案

空间形成过程
The Process of the Space Formation

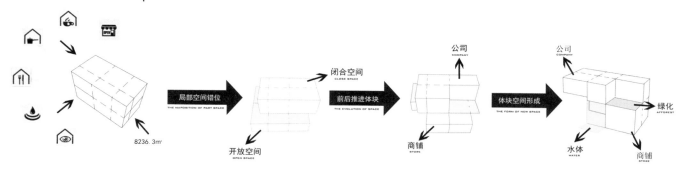

空间演变分析
The Space Evolution Analysis

建筑竖向分析
Buildings Vertical Analysis

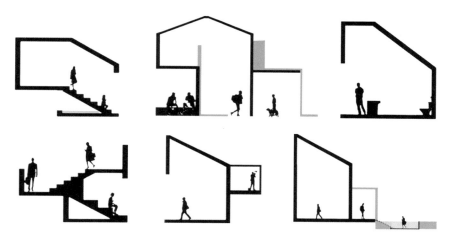

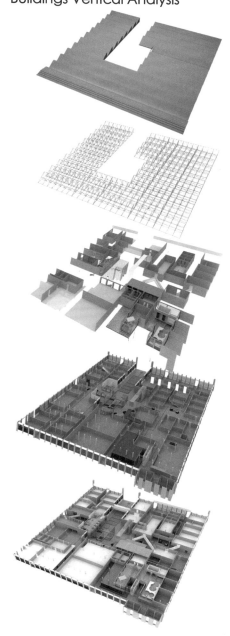

　　盒子，是限定空间却不生硬隔断空间的一个很好工具。首先我们把盒子空间解读为城市空间，将提供不同审视的角度。恰似南京传统的胡同空间，也相应留出一些狭小的空间以丰富使用者的体验和感受。并且，通过将盒子从地面上抬起，把整个空间处理成一个多种元素交相呼应的纯净且具有神圣感的容器。

　　Box is a good tool to limit the space but not separate it stiffly. First of all, we regard the space as city space and provide different point of view to watch. Just like traditional Nanjing Hutong, we should also leave some narrow spaces to enrich users' experiences and feelings. Besides, we will lift the box from the ground and making the whole space as a pure and holy container with several elements echo with each other.

过程方案 Process Plan

CIID "室内设计 6+1" 2015（第三届）校企联合毕业设计
CIID "Interior Design 6+1" 2015(Third Scssion)University and Enterprise Joint in Graduation Design

北京建筑大学
Beijing University of Civil Engineering and Architecture

小组成员：邰嘉琪 Xi Jiaqi　　姜　帅 Jiang Shuai
　　　　　刘凯南 Liu Kainan　顾文博 Gu Wenbo
　　　　　付晨辉 Fu Chenhui　王翌伟 Wang Zhaowei
　　　　　余祺盈 Yu Qiying

中国传统太极与双核理念
China's Traditional Tai chi and the Idea of Double Cores

千百年的城市发展史告诉我们，南京虽兴衰更替却生生不息，这就是南京特有的精神。而从发展格局来看，南京城墙文化浓厚，地理环境优越；近两年以南站、雨花台为核心的外城区发展迅速。在未来，南京势必形成双核城市发展体系。

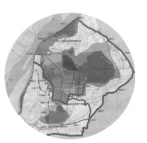

The development history of thousand years tells us Nanjing goes on with ups and downs, and this is a special spirit of Nanjing. In the view of development pattern, Nanjing is rich in city wall culture and is superior in geographical condition; recent years the outskirts develop fast with Nanzhan and Terrace of Raining Flowers as the center. In the future, Nanjing would have a double cores development system.

我们的项目正处于新核之内。而这种双核联动的发展正符合中国传统太极生生不息、相互转化的平衡之态。综合各方面，我们提取了三个元素，归纳出设计的指导思想。

Our project is from the new core. This double cores development is in accordance with the equilibrium of continuous reproduction and reciprocal transformation in Chinese Tai chi. Integrated with every side; we abstracted 3 elements and summarized the guidelines.

 从十从口，识前言者。它是一种文化，文明是对过往的传承。
Gu– means "culture", and culture is an inheritance of the past.

 从土从成，成意"百分之百"完整，土指阜堆，完全用土垒筑的围墙。
Cheng– means "complete". It's a wall made of soil.

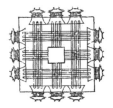

太极　　　中国传统城市设计　　　枯山水

 一种蓄藏，一种回转，也是一种平衡。
Yun–means hide rotation and also balance.

Process Plan 过程方案

1865 影工厂方案概念生成
The Concept of 1865 Cinema Factory Plan Generated

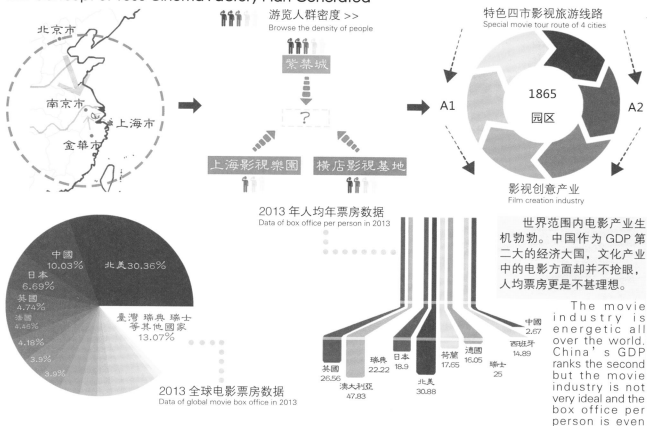

世界范围内电影产业生机勃勃。中国作为GDP第二大的经济大国，文化产业中的电影方面却并不抢眼，人均票房更是不甚理想。

The movie industry is energetic all over the world. China's GDP ranks the second but the movie industry is not very ideal and the box office per person is even worse.

通过相关产业数据分析，可以看出中国电影市场可谓前景巨大。Through the data analysis, we can see the great prospect of China's market.

园区由原晨光机器厂厂房改建而成，占地面积21公顷，总建筑面积10万平方米。
The park is reconstructed from the former Chenguang machinery factory plant. It covers an area of 21 hectares and the building area is 100,000 square meters.

CIID "室内设计 6+1" 2015（第三届）校企联合毕业设计
CIID "Interior Design 6+1" 2015(Third Scssion)University and Enterprise Joint in Graduation Design

南京艺术学院
Nanjing University of the Arts

小组成员：陆　路 Lu Lu　　　　黄文进 Huang Wenjin
　　　　刘曼羽 Liu Manyu　　　蔡文镇 Cai Wenzhen
　　　　陈飞宇 Chen Feiyu　　　许　超 Xu Chao
　　　　杨广智 Yang Guangzhi　刘佳俊 Liu Jiajun
　　　　杨　鸣 Yang Ming

历史分析
Historical Analysis

1865年五月，李鸿章筹建金陵制造局

1937年7月抗战爆发，兵工厂被迫迁往重庆，原址被日军占领

1946年9月，国民党改称六〇兵工厂

1949年4月，南京解放，工厂改属第三野战军管辖

1957年4月，采用国营晨光机器厂作为第二厂名

20007年5月，组成南京晨光1865创意产业园

1929年6月，金陵机器局改名金陵兵工厂

1945年8月，日本投降，工厂重新迁回南京

1948年12月，六〇兵工厂搬迁至台湾高雄

1952年12月，军械总厂与山西长治三〇七厂合并

1980年3月，改称南京晨光机器厂

1996年6月，正式组建晨光集团，更名为南京晨光集团有限责任公司

建筑分化分析
Analysis of the Building Differentiation

1. 园区前身为1865年创建的金陵制造局，是中国近代工业、兵器和航天工业的发祥地。
2. 园区拥有150年灿烂的近代工业文化，被称为近代"民族军事工业的摇篮"。
3. A2建筑始建于1934年，为单层钢混结构，属民国建筑。
4. 建筑大量开窗，同时避免直射光对加工制造的影响。
5. 建筑本身为包豪斯建筑风格，强调技术与艺术的新统一。

1.As the precursor of the park, Jin Ling Manufacturing Bureau was founded in 1865, which is the birthplace of China's modern industry, weapons and aerospace industry.
2.With 150 years history of brilliant modern industrial culture, the park is known as "the cradle of modern national military industry".
3.Building A2, built in 1934, is a single floor steel-concrete structure, belonging to the Republic of China building.
4.The building has a lot of windows, but avoiding the impact of direct light on the processing and manufacturing.
5. As the Bauhaus architectural style, the building shows the harmony of technology and art.

定位为南京的艺术地标，集艺术鉴赏高度与艺术门类广度为一的综合服务体。可为品牌发布会、文化艺术节、交流会、影化等提供场所。

Brand Positioning: Art landmark of Nanjing. A comprehensive service composition in which art appreciated and categories combined. It integrates the services with brand patty conference, culture and art festival, meeting, video culture, etc.

过程方案 Process Plan

自然光分析
Natural Light Analysis

以建筑受光部位的强弱作为空间划分的重要依据，结合柱网进行合理设计。把办公区与展区明显区分，把受光面较强的东面作为出口，让整个展厅上与下、明与暗、动与静有了较大的对比，不仅丰富了展示空间层次，也极大地提高展示的趣味性和互动性。

According to the light intensity, we divided the building and designed it with the consideration of column mesh. To distinguish the office area and exhibition area obviously, we set the high light intensity east as export. Making the entire exhibition has an obvious contrast in high and low places, bright and dark regions, and dynamic and static areas. It is not only enriched the display of space level, but also increased interest and interaction greatly.

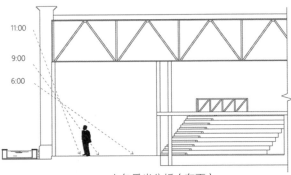

上午受光分析（东面）
Light analysis of the morning (East)

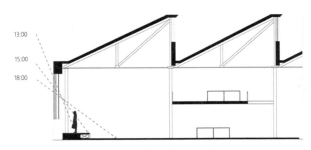

下午受光分析（南面）
Light Analysis of the afternoon (South)

People are the key factor of success: three stage strike tactics is a rapid, sustained and effective firing method before automatic weapons were invented.
Exhibits: Three stage strike tactics. Three people reload the gun and firing in turn. The original intention of the invention: rapid, sustained and effective firing to show the features of single shot weapons. Introduce the urgent need for automatic weapons.

People and weapons are the key factors of success: the invention of semi-automatic weapons. Three stage strike tactics lose its effectiveness. It's no longer a daydream to pit one against ten.
Exhibits: The invention of semi-automatic gun. Gatling gun: 200 shot / min. The original intention of the invention: one soldier against one company. They use high rapid shooting weapons to replace soldiers on battlefields and save ten thousands lives. Through Battle of Pyongyang, it will show the significant damage of semi-automatic gun.

Weapons are the key factors of success: the invention of automatic weapons, nightmare of war, the rise of the murder weapon.
Exhibits: the invention of automatic weapons. Maxim gun: 600 shot / min. The original intention of the invention: high efficiency murdering weapon. Through the Anti-Japanese War and the use of Min 24 machine guns to show the super appalling rate of fire and power of automatic weapons.

The smiling face of the Buddha made the audience back from the pain of war to peaceful reality. The beauty of life will be engraved on the audience's minds.
Exhibits: After the war, in 1865, the Jin Ling Manufacturing Bureau changed from military use to civilian use (industry products like cast copper Buddha).

文案设计
Copy design

一展厅
馆名：1：1（短兵相接）
Exhibition Hall I
Name: 1:1 (Hand-to-Hand Combat)

👤 : 👤　杀戮追求效率
　　　　Massacre pursuit efficiency

人是胜利的关键因素：三段击战术，这种射击方法在连发武器未发明时是一种快速持续有效的射击方法。
展示内容：三段击：三人交替装弹、开火。
发明初衷：快速持续有效的射击以此来体现单发武器的特性，引入对自动武器的迫切需求。

二展厅
馆名：1：10（以一当十）
Exhibition Hall II
Name: 1:10 (Pit One against Ten)

👤 : 👥👥👥👥👥　杀戮追求效率
　　　　10　　Massacre pursuit efficiency

人与武器是胜利的关键因素：半自动武器的发明，三段级战术的失效以一当十不再是空想。
展示内容：半自动机枪的发明：加特林机枪 200 发/分钟。
发明初衷：一个士兵顶上一个连，以高射速武器代替战场上的士兵，拯救数以万计的生命，通过平壤战役，突出半自动机枪以一当十的强大杀力。

三展厅
馆名：1：100（决胜千里）
Exhibition Hall III
Name: 1:100 (Victory a Thousand Miles Away)

👤 : 👥👥👥👥👥👥　杀戮追求效率
　　　　N　　Massacre pursuit efficiency

武器是胜利的关键因素：全自动武器的发明，战争的噩梦，杀人利器的崛起。
展示内容：全自动机枪的发明，马克沁机枪 600 发/分钟。
发明初衷：高效杀人武器通过抗日战争，"民 24 式机枪"的使用突显全自动武器的超强射速和骇人威力。

四展厅
馆名：0（战以止战）
Exhibition Hall IV
Name: 0 (Battle to Prevent Battles)

🔫　杀戮回归和平
　　Massacre back to peace

以佛的笑脸把观众从战争的痛苦中拉回到祥和的现实，把美好烙印在人们的内心深处。
展示内容：战争之后，1865 金陵机器制造局军转民（铸铜佛等工业产品）。

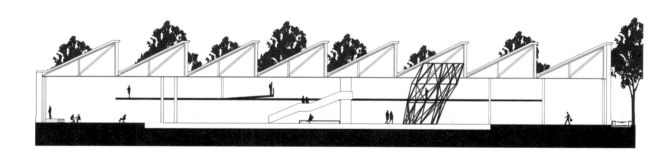

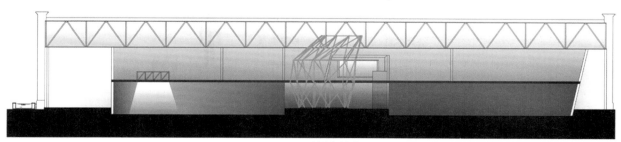

空间光线明暗分析
Spatial light tone analysis

过程方案 Process Plan

CIID "室内设计 6+1" 2015(第三届)校企联合毕业设计
CIID "Interior Design 6+1" 2015(Third Scssion)University and Enterprise Joint in Graduation Design

浙江工业大学
Zhejiang University of Technology

小组成员：张泽浩 Zhang Zehao　　夏　欣 Xia Xin
　　　　　马凯杰 Ma Kaijie　　　　孙　莹 Sun Ying

方案过程：融·生
Scheme Process : Convergence·Generation

概念提出
Concept Presentation

设计以"融·生"为主题。融，指的是历史建筑与现代空间的融合，历史建筑与场地的融合，历史建筑与环境的融合，历史建筑与城市的融合。生，指的是城市的发展交通的延伸，文脉的延续，建筑的延续，生态的延续。生态、生长、生生不息……

The design takes "Convergence·Generation" as the theme. Convergence means the integration of historical architectures and modern spaces, historical architectures and sites, historical architectures and environment, as well as historical architectures and cities. Generation means the extension of the development of cities, the city contexts, the architectures, and the ecology. Ecology, growth, circle of life…

使用者分析
Users Analysis

所有人	优美环境	Beautiful Environm
居民 Residents — 成年人	休闲娱乐	Leisure Entertainm
居民 Residents — 老人	无障碍	Barrier Free
居民 Residents — 儿童	游戏教育	Game Education
游客 Visitors — 食	特色餐饮	Characteristic Din
游客 Visitors — 住	舒适个性	Comfort Individua
游客 Visitors — 行	便捷安全	Convenient Safet
游客 Visitors — 游	文化体验	Culture Experienc
游客 Visitors — 购	特色产品	Characteristic Pro
游客 Visitors — 玩	趣味活动	Fun Activities

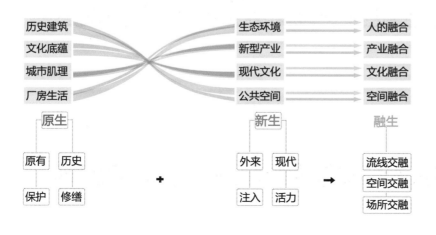

方案过程：时·镜
Scheme Process : Time•space

概念引入
Concept Introduction

弹性空间：弹性是一个物理学上的概念，引申至环境艺术设计领域，弹性设计是指在设计的过程中把未来的不确定因素考虑纳入到设计体系中，能够满足多样和变化需求的可持续发展的空间设计，设计为以后环境的发展留有余地，而非一次性将空间进行定位和设计。

Elastic space：Elastane is a concept of physics. The elastic design extended to environmental art design field means bringing the uncertain factors in the future into the design system during the process of designing, and the space design of sustainable development which can satisfy the needs of diversity and change. The designs leave room for the development of environment in the future, rather than position or design the space for once.

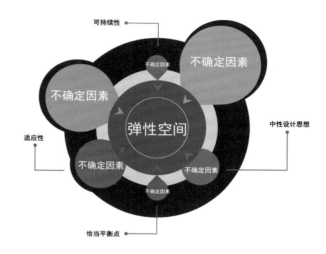

设计意向与思路
Design Intention & Idea

（1）满足不同使用者的个性需求，解决困扰我们的空间使用问题。
（2）能够应对时间的变化，从使用者的角度以适应时间变化作为设计的核心理念。
（3）增加使用者的参与度，发挥设计者和使用者两方面的积极性。

(1) To content individual needs of different users. And to solve the space-use problem troubling us.
(2) To be able to cope with the changes of time. Take adapting to the changes of time as the core concept of design.
(3) Increase the participation of the users. Bring the initiative of designers and users into play.

设计概念构建
Design Concept Establishment

（1）功能混合
（1）Functions Mix

（2）引入"自助"方式
（2）Bring in the Method of "Self-Service"

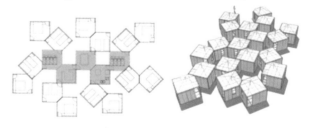

（3）可拆卸结构的使用
（3）The Use of Removable Structure

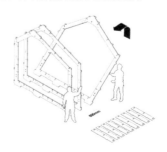

CIID"室内设计 6+1"2015（第三届）校企联合毕业设计
CIID "Interior Design 6+1" 2015(Third Scssion)University and Enterprise Joint in Graduation Design

工业遗产保护与室内设计
Industrial Heritage Protection and Interior Design

工业遗产保护与室内设计 Industrial Heritage Protection and Interior Design

CIID "室内设计 6+1" 2015(第三届)校企联合毕业设计
CIID "Interior Design 6+1" 2015(Third Scssion)University and Enterprise Joint in Graduation Design

一等奖
First Prize

金陵印
Jinling Seal

高　　校：	同济大学
College:	Tongji University
学　　生：	马潇潇　张黎婷
Students:	Ma Xiaoxiao　Zhang Liting
指导教师：	左琰
Instructors:	Zuo Yan

关联页：66-73

Industrial Heritage Protection and Interior Design
工业遗产保护与室内设计

马潇潇 Ma Xiaoxiao
张黎婷 Zhang Liting

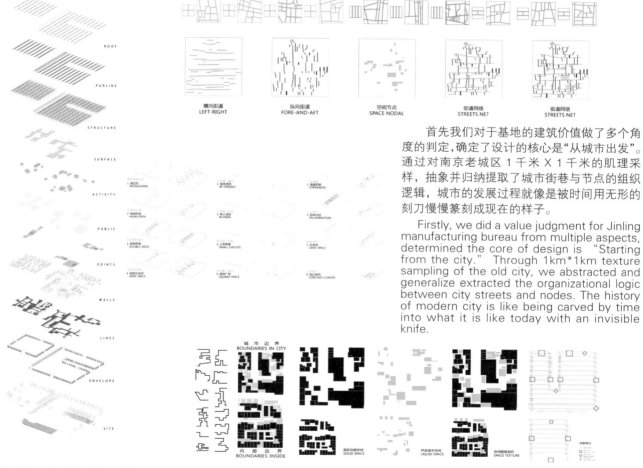

首先我们对于基地的建筑价值做了多个角度的判定,确定了设计的核心是"从城市出发"。通过对南京老城区1千米X1千米的肌理采样,抽象并归纳提取了城市街巷与节点的组织逻辑,城市的发展过程就像是被时间用无形的刻刀慢慢篆刻成现在的样子。

Firstly, we did a value judgment for Jinling manufacturing bureau from multiple aspects, determined the core of design is "Starting from the city." Through 1km*1km texture sampling of the old city, we abstracted and generalize extracted the organizational logic between city streets and nodes. The history of modern city is like being carved by time into what it is like today with an invisible knife.

进一步我们研究了城市中街巷人类行为方式与空间的对应关系,发现城市空间总是不断在向着"适应人们"的新目标。如何将这样的理解篆刻到基地中完成一个有价值的设计正是我们想要通过这次的毕设探索的问题。

Then we studied the correspondence between human behavior and city space, and found that city space always changes, following the ways people live. How to complete a valuable design with this understanding is what we want to solve in this graduation design.

在这次的设计中,我们希望最终能够达到反映地域文化、延续城市脉络、保留兵工记忆、激发街区活力的四个目标。

In this design, we attempt to achieve four goals: to reflect regional culture, to continue the city sequence, to save the ordnance memory, and to stimulate vitality of neighborhood.

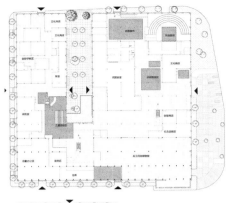
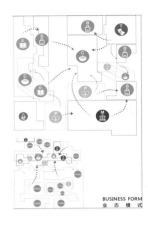
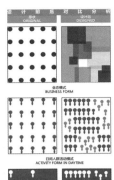

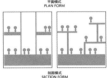

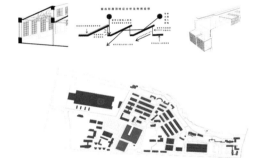
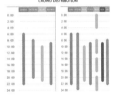

　　从城市出发的设计本质其实是对于城市中人的活动与空间关系的探索，我们的室内空间设计也试图用同样的方式篆刻出契合室内活动的空间尺度，并最大程度地对金陵制造局的厂房立面、结构进行保留。相对于城市中的多样活动方式与多层次的空间模式，室内常见的通过式公共空间往往会打断人们的活动体验，因此设计中我们丰富了建筑内部的界面，模拟街巷空间中邻里关系，自然地引发交流的欲望，拉开公共空间层次，试图提供更多的活动和交流的可能性，让人们在城市中养成的活动习惯在新的场所和功能中，依然能得以匹配和契合。

　　The essence of "starting from the city" is to explore the relationship between human behavior and city space. In Our interior design, we try to work out the spatial scale which fits into indoor activities, and mostly retain the plant facade, the structure of Jinling Manufacturing Bureau to the greatest extent. Compared with the city's various activities and multi-level spatial pattern, public space which is commonly used in indoor designing will usually interrupt people's activities. So in our design, we have enriched the interface inside the building, stimulate streets neighborhood, naturally trigger desire to communicate, pull ahead public space level and try to provide more chances and activities for people to communicate, so that people are able to adapt in a new atmosphere with their habit which developed in cities.

　　城市进化的关键过程就是对于A1、A2合理利用模式。这个模式需要改善建筑原本的不足并且激发建筑新活力。在通过系统分析之后，我们决定将原本比较单一且各自为营的业态模式进行改变，变成多元化的、互相拉动的业态模式，且将日间和夜间吸引的人气进行优化，丰富活动的内容，拓展建筑服务的时间及范围，划分出更多层次的空间从而改善人群的分布状态。这些改变都将是建立在对建筑原本特质的思考和对南京人的活动方式产生的结果，对于我们的设计来说，室内人们的行为模式将不再仅仅受限于一个或两个房间当中，也不会固步自封于建筑的外壳当中，而是与整个建筑一起成为生命共同体，变成人们在城市中行为模式的一部分。

　　The key process of city evolution is the rational use of A1 and A2. This mode needs to improve the original shortage and stimulate new vitality of the building. After system analysis, we decide to change the single, original and respective business model into a diversified and mutually stimulating business model. At the same time we'll optimize the amount of people in daytime and nighttime, enrich the content of activities, expand building service time and scope, divide more levels of space to improve distribution of population. All these changes will be based on the thinking of the original building's characteristics and the results of people's activities in Nanjing. In our design, indoor activities will no longer be limited to one or two rooms, neither be stuck in housing construction, but become a community life with the entire building, also a part of people's life in the city.

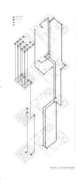

Industrial Heritage Protection and Interior Design 工业遗产保护与室内设计

重点空间深化
Deepen the Key Space

对于场所中的不同人群，我们也进行了其各自的活动模式的设计，试图创造一个全龄化、全时化的空间。除了对于整体空间的组织把握，我们还针对四个重点空间进行了设计的深化。这四个空间分别是民国食堂、全龄学堂、实验剧场和第三场所。

For different groups of people, we also carry out their own activity design. We try to create a full-age and full-time space. In addition to hold the whole space organization, we also deepen our design in four key spaces which are: Republic Canteen, Full Age School, Experimental Theater and the third workplace.

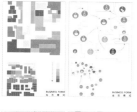

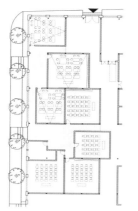

 大报告厅模式 LECTURE MODE
 工作坊模式 WORKSHOP MODE
 小型报告厅模式 SMALL LECTURE MODE
 学术沙龙模式 SALON MODE
 长桌模式 TABLES IN LINES
 双人桌模式 DOUBLES
 小组模式 GROUPS MODE
 单人桌模式 SINGLE

在整个空间当中，我们使用木材、青砖、白墙加上彩色的点缀，形成空间整体文雅而明亮的氛围。同时，把传统的建筑质感和现代语言碰撞在一起，形成一套完整的界面语言逻辑。明亮的基调可以改善、优化原本工厂沉闷的氛围，而材质的比重转换又可以产生不同空间特质，使其与不同的功能相契合。

In the whole space, we use timber, brick, white walls and colorful embellishment to create an elegant and bright atmosphere. We also combine the traditional architectural texture with modern language to form a complete set of interface language logic. Bright tone can improve and optimize the dull atmosphere of the original factory, the proportion of material transformation can produce different spatial characteristics, and all these can fit its different functions.

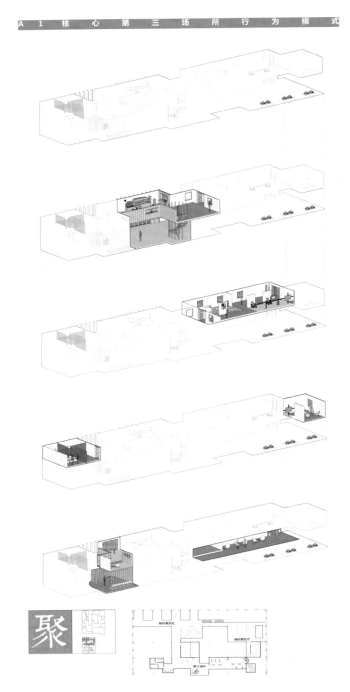

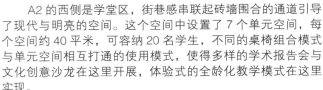

A1 第三场所效果图（首层）
A1 rendering of the third place (the first floor)

A1 第三场所效果图（上层）
A1 rendering of the third place (the upper layer)

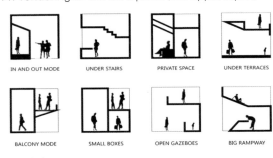

剖面活动　Section Modes

A2的西侧是学堂区，街巷感串联起砖墙围合的通道引导了现代与明亮的空间。这个空间中设置了7个单元空间，每个空间约40平米，可容纳20名学生，不同的桌椅组合模式与单元空间相互打通的使用模式，使得多样的学术报告会与文化创意沙龙在这里开展，体验式的全龄化教学模式在这里实现。

穿插在以工作室功能为主的A1空间中，为了丰富设计师的交流空间，提供灵感碰撞的可能性，我们特别设置了上下关系丰富的"第三场所"。这种对于街巷空间中邻里关系的模拟，自然地引发了交流的欲望。这里有可供休息停留的咖啡吧，结合着二层的民俗展览馆与为艺术家们提供的自由展览空间，使得大众群体在不影响艺术家们工作的情况下有机会在这里获取最新鲜的创造力。相对私密的会谈空间与流动性的主题休息区，差异化的空间尺度使得聚合来的更加简单。

School district is on the west side of A2. The sense of the street connects the channel which is surrounded by brick walls to guide the modern and bright space. The space sets up seven units, which can hold 20 students in about 40 square meters. With different sets of desks and chairs and the model of the units get through, make the diversity of the symposium and cultural creativity salon develop here, experience the whole process of teaching mode implements here.

Interspersed in the studio features based A1 space, in order to enrich the exchange space of designers and provide the possibility of inspiration, we have especially set up a "third place" of the upper and lower relations. Simulation for the relationships of neighborhood that is in the space of street, naturally leads to the desire to communicate. Here are some coffee shops, with two layers of folk art exhibition hall and the free exhibition space provided for the artists, the public group has the opportunity to get the freshest creative work without affecting the work of the artists. There is relatively private space and subject area of mobility and the spatial scale of differentiation make it easier to aggregate.

Industrial Heritage Protection and Interior Design　工业遗产保护与室内设计

食堂平面图
RESTAURANT PLAN
1:100

食

在A2庭院东侧，我们设计了一个近八百平米的，具有时尚感、人情味与主题性的民国食堂。我们将民国画报的意向与现代元素碰撞在这个特别的食堂中，暴露的钢架下平面，靠近入口的优越位置使得围桌吃饭与约会小聚都将成为可能。这里贩售的食物以南京特色的小吃与菜肴为主，不同区域的划分有效地分隔了人流。

On the eastern side of the A2 courtyard, we have designed a nearly eight hundred square meters cafeteria of the Republic of China, which has a sense of fashion, human interest and theme. We mix the republic of china style pictorial and modern elements in this special canteen, the exposed steel frame under the plane, located near the entrance of the dinner table and the date together will become possible. The food here is mainly special Nanjing snacks and dishes. Different regions separate the stream of people effectively.

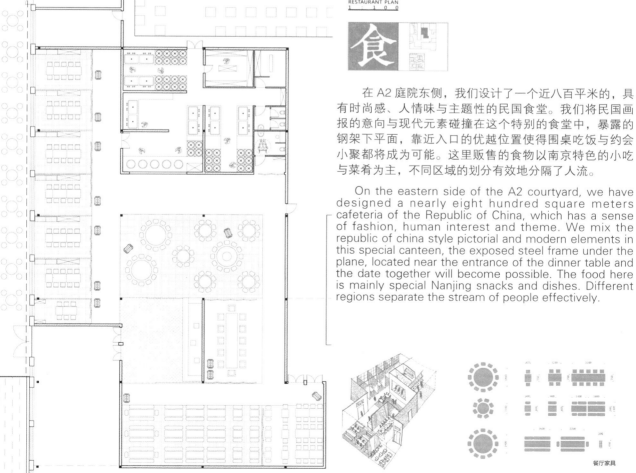

餐厅家具

071

工业遗产保护与室内设计　Industrial Heritage Protection and Interior Design

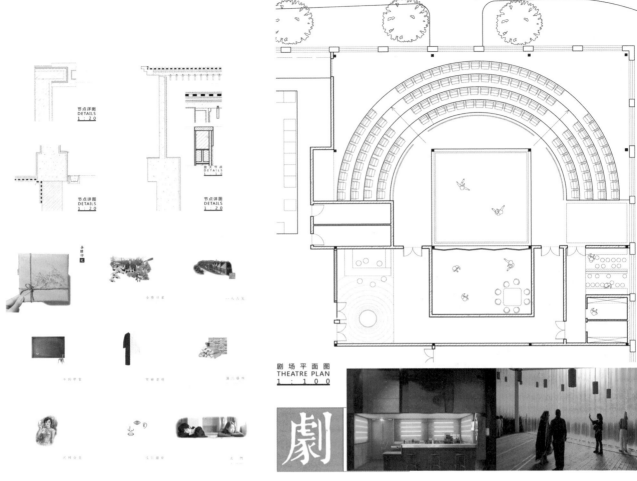

正邻着民国食堂东北侧的是小型的实验剧场，侧重于年轻人消费群体的剧场在空间氛围上保持工业感并以红色调点缀。屋顶上可电动开合天窗与幕布配合灯光营造不同的演出氛围，与后台相接的可变舞台为现代话剧与小型公演提供了更多可能性。

Adjacent to the northeast side of the cafeteria is a small experimental theater which focuses on young consumers. Its space maintains a sense of industrial and red color in decoration. On the roof, the retractable skylight and curtain coordinated with the lighting to create different atmospheres. The variable stage connected with the backstage provides more possibilities of modern dramas and small openings.

常规图纸之外，我们还做了延伸的设计——"金陵印纪"，成为此次设计的过程记录、再表达与感悟的汇集。一套共分为八册，分别阐述了城市、兵工、传统与现代语言在设计上的矛盾冲突与相交相融，以及对于此次工业遗产保护设计的体会。

Besides the conventional drawings, we did some extended design as well— "Jinling Seal", which becomes the collection of the recording of process, the expression and the comprehension. A set is divided into eight volumes, respectively elaborated the conflict and blending intersection of the city, ordnance, traditional and modern languages in design, and the understandings of the industrial heritage protection design.

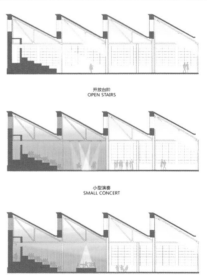

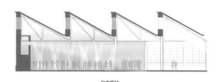

实验剧场的不同使用模式

学生感想
Students' Thoughts

　　三月含笑花开，我们来到了南京。不管曾经来过多少次，拥有多少关于南京的记忆，因为这次的毕业设计，整个城市在我们眼中都是全新的。它的繁荣、平凡、尊贵、市井、祥和、伤痛，都在设计的过程中反复敲击着我们，也让我们感到迷茫。于是，我们决定，试图用城市的眼睛看这个城市，一如用自己的眼睛看自己那样。现在，已是六月，是茉莉花开的日子，借这个毕业设计表达我们对这城市、这建筑、这些人们的理解与祝福。谢谢南京，谢谢1865。

　　We came to Nanjing in March when the flowers bloomed. Whatever how many times we came or how much memory we had, we felt the whole city was new for us because of the graduation projects. Its prosperity, common, dignity, and philistine, harmony, as well as pain impacted us again and again during the designing process. Meanwhile, it also let us feels confused. Therefore, we decided to try to have a look at the city by eyes of the city, just like we look at ourselves by our own eyes. Now it is June, the days of blooming jasmine flowers. We are going to express our understandings and blessings to the city, the architectures and the people through the design. Thank you, Nanjing. Thank you, 1865.

It brings the connection of the context to the inner space, trying to build a space mode that corresponds to the activity of the people who live in the city for a long time, and meanwhile contains new content. The concept is clear, the thought is precise, and the logic is distinct. With strong site awareness, the function division and streamline can notice the current situation around the building. Building the partial interlayer space does not only add the effective building area, but also strengthen the space humanized scale. However the design for upper & lower interface division and space relation of the added part is inconsiderate, so it needs to be improved.

——Jiang Wenyi

专家点评
Experts' Comments

　　设计团队从项目所在地的城市肌理入手，将对城市历史街巷的印象带入室内设计。创意创业园的构想成熟，功能空间分配得当，动线流畅，有节奏感。以"剧，食，聚，学"等复合式的服务业态，穿插其中，模拟效果有很好的可实施性。时间在项目中的表达充分，使整个空间的设计多出了一种源于生活的深意，金陵机器制造局这座历史建筑由此形成新的金陵印象。

——陈卫新

教师点评
Teachers' Comments

　　设计者能站在城市的视角，关注城市、空间、人的活动彼此之间的关系。方案结合城市文脉，力求保留厂区的兵工记忆。从城市脉络肌理提炼出设计的理念，城市脉络的关系引入到室内空间中，尝试营造出在一个能与城市中长期生活的人们活动契合的同时又拥有新的内容的空间模式。概念明确，思维严谨，逻辑清晰，功能分区与流线组织能关注到建筑周边的现状条件，有很强的场地意识。局部夹层空间的加建，既增加了有效的建筑面积，同时更加强化了空间人性化的尺度，但加建部分上下界面划分及空间关系设计考虑不周，有待完善。

——姜文艺

　　In the view of the city, the designer could focus on the relations between the city, its space and human activity. Combined with the city's unity and coherence in writing, the project tries to keep the memory of the ordnance in the factory area. The concept of the design was created on the basis of the feature and the context of the city.

The design team started the work with the context of the located city. They will bring the impression of the city's historical street to the interior design. The thought of the originality innovation park is mature, its functional space allocation is reasonable, and the kinetonema is fluent and rhythmic. The project inserts the complex service commercial activities "opera, food, gathering, learning". The result shows that it's worth implementing. The time is fully expressed in the project. It gives the design of the whole space a profound meaning from our lives. Then, the building, Jinling Machinery Manufacturing Bureau, becomes the new symbol of Jinling.

——Chen Weixin

工业遗产保护与室内设计 Industrial Heritage Protection and Interior Design

CIID "室内设计 6+1" 2015（第三届）校企联合毕业设计
CIID "Interior Design 6+1" 2015(Third Scssion)University and Enterprise Joint in Graduation Design

一等奖
First Prize

基质的再生
Matrix

高　　校：	华南理工大学
College:	South China University of Technology
学　　生：	马仓越　廖绮琳　俞快
Students:	Ma Cangyue　Liao Qilin　Yu Kuai
指导教师：	姜文艺　陈建华
Instructors:	Jiang Wenyi　Chen Jianhua

关联页：74-81

马仓越 Ma Cangyue　　廖绮琳 Liao Qilin　　俞快 Yu Kuai

概念生成
Concept Generation

在设计过程中，我们思考如何在富有历史感的旧厂房中，保留原有文化片段的同时又能注入新的活力，将新与旧的冲击直接演绎出来而又达到一种平衡。于是，我们需要在厂房内加入一种新的元素以激活厂房。

新加入的元素必须对旧厂房各部分构件不产生巨大的影响，并且尊重原有建筑的特征和限制，而作为一种全新的、多元的基质，承载新的功能和内容，创造新的空间对话。

我们将这种新加入的系统称为"MATRIX"。

During the designing, we consider how to introduce dynamic elements into the old factory building with a sense of history while retaining the original cultural fragments, showing directly the impact between the new and the old and achieving a balance. So we need to bring in a new element to activate the old factory building.

The new element should have no significant impact to each part of the old factory building, respects the original architectural features and restrictions. It needs to bear the new features and content, to create a new space interaction as a new and multivariate matrix.

We call this new system "the MATRIX".

MATRIX 是什么？
What is the MATRIX ?

（1）.MATRIX 是一种限制。

（1）. The MATRIX is a kind of limit.

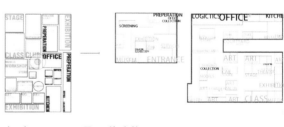

（2）.MATRIX 是一种功能。

（2）. The MATRIX is a kind of function.

 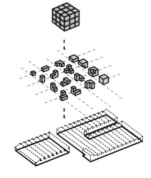

（3）.MATRIX 是一种形态。

（3.）The MATRIX is a kind of form.

（4）.MATRIX 是一种精神。

（4）. The MATRIX is a kind of spirit.

1. MATRIX 是一种限制。厂房原本的柱网密集、层高低、进深大，这对于新加入的元素本身就成为了一种制衡，新加入的元素需要尊重原本的母体结构并且适应其发展。

2. MATRIX 是一种功能。新加入的 MATRIX 包含了各种不同的功能，包括兵器博物馆、艺术家工作坊、艺术沙龙、实验剧场等，新的功能带来了新的活动，吸引人们到此处进行不同的交流，迸发新的思想。

3. MATRIX 是一种形态。具体化成不同形态的体块，放置到场地内部，充分考虑因素以及内部相关性，构成交通流线、物理环境等外部新的功能分区。

4. MATRIX 是一种精神。在场地悠久的历史中，金陵制造局作为近代中国兵工历史的先河，发明制造了属于中国的兵器，是当时中国人创造力的结晶和见证。如今，场地中引入了新的艺术产业，人们在这里汲取了古老的厂房所承载的历史的创新精神，这将赋予艺术工作者们更多的灵感，为中国未来创造力提供无限的可能性。

1. The MATRIX is a kind of limit. The old factory building enjoys dense column grid, short floor height and big chin deep, which is a kind of checks and balances for the new element. The new element should respect the original matrix structure and adapt to its development.

2. The MATRIX is a kind of function. The new MATRIX contains a variety of different functions, including armory museum, artist workshop, art salon, experimental theater, etc. New functions bring new activities, which attract people to communicate here for new ideas.

3. The MATRIX is a kind of form. We put different forms of massing which is made concretely in the site to constitute a new function division, fully considering the external factors such as traffic stream line and physical environment, as well as the internal correlativity.

4. The MATRIX is a kind of spirit. In the long history of site, as the first time in military history of modern China, QuanLing Arsenal invented and manufactured China's own weapons, which is the crystallization and witness of Chinese creativity. Now new art industry is brought into the site, from where people absorb the historical pioneering spirit of the old factory building. This will inspire more inspiration to artists, and provide infinite possibilities for China's future creativity.

概念与空间分析
Analysis of Concept and Space

A1 和 A2 平行排列，与金陵制造局的牌坊半围合出一个矩形的空地。综合考虑城市交通、园区车行和人行系统，本方案将牌坊重新利用作为整个园区的人行入口，将矩形空地作为进入 A1、A2 厂房前的前广场，并利用 A1 的大门作为两个厂房的主入口，以此活化建筑周边的空间，将建筑作为园区与公众之间的交流平台。

Factory building A1 and A2 are parallel, makes a semi-enclosed rectangular clearing with the memorial archway of QuanLing Arsenal. Synthetically considering the urban traffic, driving system and walking system, this scheme reuses the memorial archway as the park entrance, and uses the rectangular clearing as forecourt of factory building A1 and A2. It also uses the gate of A1 as the main entrance of them, activating the space around the buildings, making the buildings a communication platform between the park and the public.

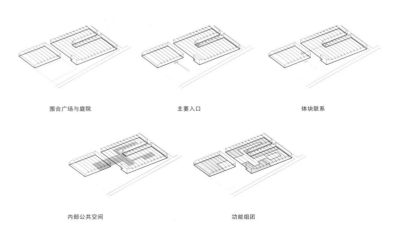

围合广场与庭院　　　主要入口　　　体块联系

内部公共空间　　　功能组团

平面分析
Plane Analysis

（1）. 功能分析
(1). Fuction Analysis

（2）. 流线分析
(2). Streamline Analysis

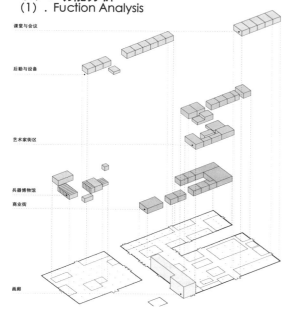

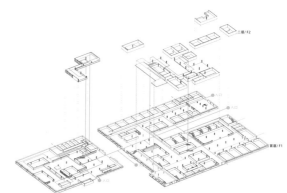

（3）. 图底分析
(3). Figure – Ground Analysis

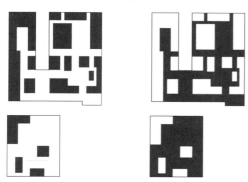

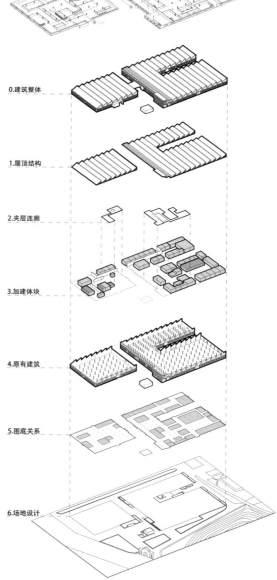

　　插入的功能体块分隔了厂房的空间，留出的空间作为介于MATRIX体块和厂房间的流动空间，不仅起到了联系交通的作用，也成为了各个功能延伸叠加的区域，这些空间不被定义功能，可随着不同的活动和需求变化，形成了更加灵活的体系。

　　The inserted function massing divides the space of the factory building. The spaces we leave, between the massing and the factory building, not only have the function of communications but also become the superimposed area of each function. These spaces are not function defined, they can be changed with different activities and demands, form a more flexible architecture.

（4）. 构成分析
(4). Component Analysis

　　我们认为加建、改建等改造行为是建立在原有建筑A1和A2之上的。我们可以将A1、A2视为一种原始的带矩阵的母体，在尊重原有基质的基础上再加入带特殊功能的多元载体。从一种基质发展成为多元载体——即MATRIX的本意，在这个偌大的厂房里面，完成新旧两次革命的对话。

　　We think the transformation behavior such as construction and reconstruction is built on the original building A1 and A2, which we can treat as a primitive matrix with matrices. Multidimensional carrier with special function based on oriental building. Developed into multidimensional carrier from a kind of matrix——original meaning of the MATRIX, to complete the interaction between the two revolutions of the new and the old in this huge factory building.

工业遗产保护与室内设计 Industrial Heritage Protection and Interior Design

主要公共空间
The Main Public Space

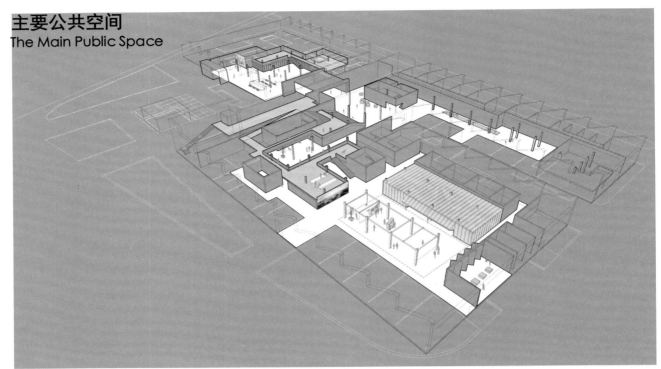

四个空间节点
Four Space Nodes

博物馆 | Museum

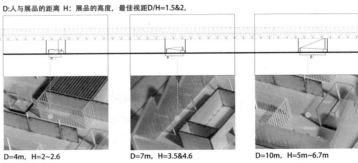

D=4m, H=2~2.6 　　D=7m, H=3.5&4.6 　　D=10m, H=5m~6.7m

-外部平台视线分析-

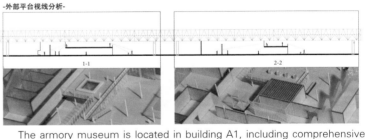

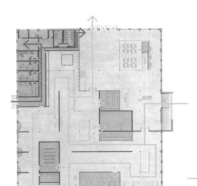

——参观流线
——后勤流线

　　兵器博物馆位于 A1 建筑内部，包括综合展览区、两个小展区、放映厅、纪念品商店等功能。为了增强博物馆的趣味性，公共空间二层平台架于博物馆之上，使得内部与外部有部分视线交流，外部的人也能略窥博物馆其中之一二。

　　艺术街区位于 A2 建筑的东南角，主要由体块以及连廊组成，包括了艺术家工作室、沙龙、咖啡厅等功能。体块与连廊形成街道亲切的感觉，这里不分外来者与本地人，只有创作者与欣赏者，打破传统的整片的艺术家街区，促进市民与艺术家的交流。体块在设计的时候充分考虑不同类型的艺术家的使用需求，建筑界面与规格都进行了一定的区分。二层廊道除了起连接作用还增强了空间的多样性。在街区的中心，有一处比较开放的宽敞空间，为各种人群的交流与学习提供了更好的平台。

　　The armory museum is located in building A1, including comprehensive exhibition area, two small exhibition areas, a projection, a souvenir shop and other functions. The second floor of the public space is on top of the museum, in order to increase the interest of the museum. So there would be line of sight communication between the internal and external part, and people outside can have a brief glimpse of inside.

艺术街区 Art District

　　Art district is located at the southeast of building A2, which is formed of massing and corridors including artist studio, salon, café, and so on. The streets formed of massing and corridors make people feel warm. There are no outsider or native, only the creator and the audience. We break the traditional entire art district to promote the communication between citizens and artists. Different kinds of artists' demands were fully considered when we designed the massing, and architectural interface and specification have also been distinguished. Corridors in the second floor play a role of connecting and enhancing spatial diversity. There is a relatively open spacious space at the center of the block, which provides a better platform for all to learn and exchange.

Industrial Heritage Protection and Interior Design 工业遗产保护与室内设计

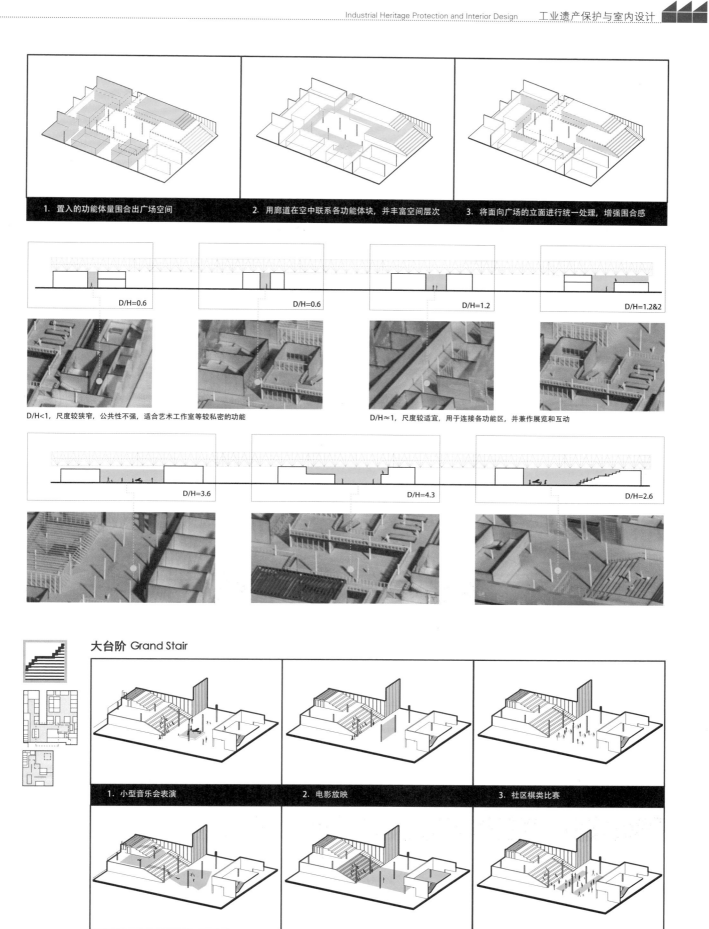

在建筑A2通往建筑A1的入口处，有一个四周界面都不一样的过厅；无论从庭院到风雨廊到室外，还是从艺术街区到特色书店到商业街区，都能看到这里标志性的大台阶。每一级台阶的高度都设置成适宜休憩的400毫米；平时，这里是整个文化体验中心的人都可以享用的休息平台；当遇上活动的时候，过厅可以适当设置简易放映装置或者小舞台，大台阶就自然变成了阶梯座位。

在建筑A2东北部有一处开放的室内广场空间，毗邻于旁边的就是实验剧场。以当代实验先锋喜剧演出为主的剧场，自由度高、空间不需要过渡设计，给剧作留下更多创作的空间；舞台的概念在这里被弱化，观众可以只是席地而坐，因为更多注重的是表演者与观众之间的互动。面对广场的剧场立面可以被打开，空间的隔断被打断以后，可以结合周围的界面围合产生新的活动，例如，发布会、跳蚤市场、放映会、大型展览等。

There is a vestibule that different from interface around on the entrance of the way from building A2 to A1. Whether from the courtyard to the gallery and to outside, or from the art district to characteristic bookstore and to the shopping street, people can see the symbolic large stairs. The height of each stair is set to be 400mm, which is appropriate to rest. In normal times, the whole people of the Culture Experience Center can enjoy this platform for a rest; during big events, vestibule can be installed with simple screening device or small stage, then the large stairs naturally become stadium seating.

实验剧场 Experimental Theathre

There is an open indoor square in the northeast of building A2, abutting the experimental theatre. The shows of the theatre are mainly contemporary experimental pioneer comedy. So the freedom of the theatre is high, and space does not need transition design to leave more room to create for the librettists. The concept of the stage is weakened here, audience can only sit on the ground, pay more attention to the interaction between the performers and the audience. The theater facade that faces the square can be opened. After space partitions are broke, new activities such as release conference, flea market, screening, large exhibition and so on, may generated enclosing with the interface around.

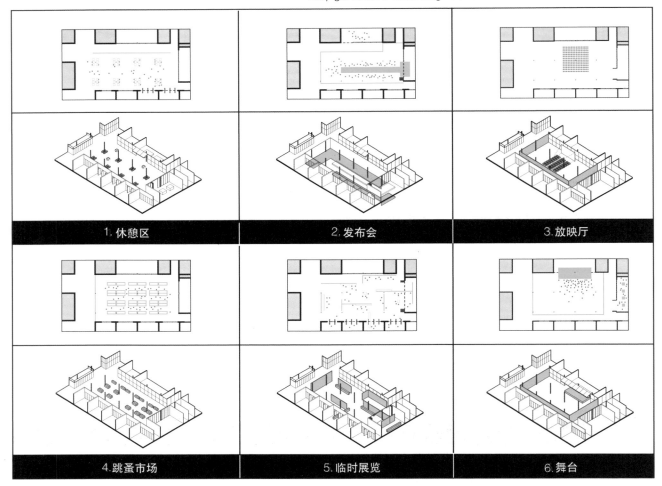

1. 休憩区　　2. 发布会　　3. 放映厅　　4. 跳蚤市场　　5. 临时展览　　6. 舞台

该区域被先锋剧场、课堂、临时展览区等公共性强的功能体块包围，围合出尺度最大的公共空间。该空间的功能不被强制定义，同时通过收纳在体块内的可移动高隔断进行灵活的空间分隔，能够为各种可能发生的活动提供场所，增强空间的可变性。

The area is surrounded by strong public function massing such as pioneer theater, classroom and temporary exhibition area, enclosing to a largest public space. The function of the space is not forcibly defined, at the same time flexibly divides space with the movable high partitions contained in the massing, for providing places for all kinds of activities and enhancing spatial variability.

学生感想
Students' Thoughts

很庆幸通过CIID联合毕设得到了与其他院校交流的机会,并且第一次接触室内设计的内容,这对于毕业设计来说有了更多的意义。旧工厂的改造更新是当今的热点问题,也是一个很有挑战性的题目。借助1865创意产业园的设计,我们不仅了解了一座城市、一段历史,也树立了对待工业遗产的态度。作为一段历史的见证者,工业遗产不仅是一栋建筑,也是一种精神,我们的设计既要保证其合理的使用功能,也要将其精神延续,感染后人。这次毕业设计让我们思考了这些问题,受益匪浅。

It's my fortune to acquire the opportunity to communicate with other colleges through the CIID Joint Graduate Project, and it's my first time to get in touch with the content about the Interior Design. It means more for the graduate project. The old factory reconstruction is a hot issue at present. It's also a challenging theme. According to the 1865 Innovative Industry Campus Project, we did not only understand a city, a period of history, but also set up an attitude toward the industry relics. As a witness of a period of history, the industry relic is not only a building, but also a kind of spirit. Our design should ensure its reasonable function. Its spirit should also be extended to inspire the later generation. This project made me think about these issues. I benefited a lot.

教师点评
Teachers' Comments

作品MATRIX,整体设计思路清晰,对场地环境和原有建筑有较深入的调研和解读;对历史建筑再利用方式有明确的态度。作品通过一个英文多义词MATRIX对设计概念进行了很好的凝练和分析,它包含了场地限制、功能载体、形态生成和场所精神4种含义。这些概念在具体设计中得到了很好的体现。设计整体完成度高,处理空间尺度、空间组织及功能流线都十分完善;通过4个模块空间把建筑中的重要节点进行重点表现,更好地解决了室内设计中空间塑造、材料选择、氛围营造、细节处理和技术措施等主要问题;作品图纸齐全、模型制作精致。完整的设计概念和充满理性的专业表达,使整个作品具备优秀毕业设计的品质。

——周立军

The work MATRIX, its general design thought is clear. The designer did a deep research and analysis, and took a clear attitude toward the way of the historical buildings reuse. According to the polysemy MATRIX, the work extracts and analyzes the design concept well. It includes four meanings: the site limitation, the function carrier, form generation, and site spirit. These concepts are well reflected in the detailed design works. The degree of design completion is high in general. The spatial scale processing, space organization and function streamline is perfect. The key point of the building is mainly expressed according to four model spaces. It's a better way to solve the main problems when designing the inner space. The problems include: space shape, material choosing, and atmosphere construction, detail processing, technique measures and so on. The drawing is completed, and the model is delicate. The completed design concept and professional express filled with rationality make the whole works have the quality of the excellent graduated design project.

——Zhou Lijun

专家点评
Experts' Comments

设计者对原厂房的基本结构以及厂区的历史文脉做了充分的调研,在此基础上,设计团队体现了对旧厂房遗存保护的理解与改造控制的能力。对办公空间设计有一定的创新性,对空间把控收放自如,功能安排合理,动线规划严谨,符合各项专业的要求,最大限度地保留了原厂房的基本构架并尽可能地把空间利用最大化,有效地增加了办公使用面积。通过设计凸显了旧厂房的特质,色彩运用大胆,有活力,新旧结合得体,黄色与黑色的搭配有时代气息,有较强的创新性。

设计组在答辩中表现出整体的专业水平与合作精神,作业工作量大,模型和图纸表达完整。

——林学明

The designer did a sufficient research on basic structure and historical context of the original factory. On the basis, the design team showed their ability to understand the protection and reconstruction control of the old factory's relics. The design of the office room is relatively creative, and the room could be controlled freely. Its function layout is reasonable. The kinetonema program is precise. It confirms to the request of each professional items. The basic structure of the original factory is remained in the maximum. The designer made the maximum use of the room as possible as he can. According to the project, the feature of the old factory is highlighted. The usage of the color is adventurous and energetic. The combination of the old factors and new factors is proper. The collocation of the black and yellow has the sense of times. It's relatively creative.

The performance of the design group reflects their general professional levels and team-work spirit. The work amount is large, the model and the drawing are completely expressed.

——Lin Xueming

工业遗产保护与室内设计 Industrial Heritage Protection and Interior Design

CIID "室内设计 6+1" 2015（第三届）校企联合毕业设计
CIID "Interior Design 6+1" 2015(Third Scssion)University and Enterprise Joint in Graduation Design

二等奖
Second Prize

交融
Blending

高　　校：	华南理工大学
College :	South China University of Technology
学　　生：	廖喆璇　周靖涵　郑宇晨
Students :	Liao Zhexuan　Zhou Jinghan　Zheng Yuchen
指导教师：	姜文艺　陈建华
Instructors :	Jiang Wenyi　Chen Jianhua

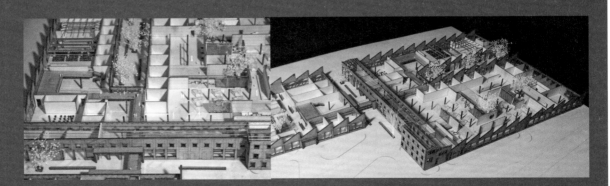

关联页：82-89

Industrial Heritage Protection and Interior Design　工业遗产保护与室内设计

廖喆璇　Liao Zhexuan
周靖涵　Zhou Jinghan
郑宇晨　Zheng Yuchen

场地区位分析
Site Location Analysis

晨光1865创业园位于南京秦淮河畔，与明城墙隔河相望，南连著名的雨花台风景区，西接金陵大报恩寺遗址，占地面积21万平方米，建筑面积10万平方米。其所在地为李鸿章于1865年兴建的金陵制造局旧址，园区内大部分建筑为民国时期的工业建筑，并且保存质量良好，如同一座近代中国工业博物馆，记录着民族工业发展的历史轨迹。

Chenguang 1865 Innovation Park is located near the Qinhuai river of Nanjing opposite City Wall of Ming Dynasty. It is south to famous Yuhuatai scenic spot and west to Jinling Dabaoen Temple historic site. It covers 210,000 square meters and has a construction area of 100,000 square meters. Its location is former site of Jinling Manufacturing Bureau constructed by Li Hongzhang on 1865. The majority buildings in the park are industrial buildings during the period of the republic of China. They have great preservation quality and look like a modern China's industry museum where a historical route of development of national industry is recorded.

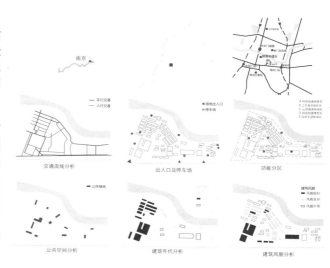

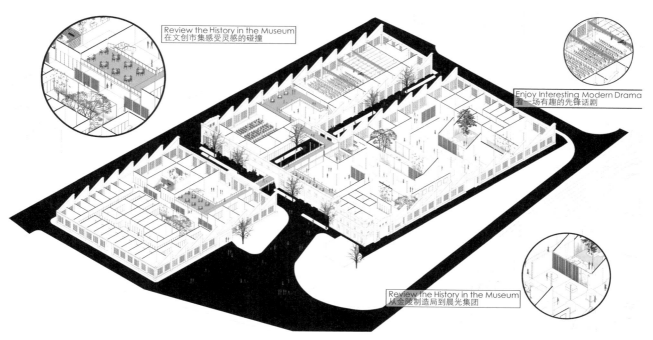

083

工业遗产保护与室内设计　Industrial Heritage Protection and Interior Design

概念生成
Concept Generation

　　南京位于江南江北的交界处，前可挹长江之水，背可倚紫金之巅，兼得山之雄水之秀，曾多次遭受兵燹之灾，亦缕缕从瓦砾荒烟中重整繁华。南京的历史就像人跌宕起伏的一生，有高潮低谷，曾是南唐盛世之金陵，也曾经历大屠杀带来的灾难。但正是所有的这些造就了今天的南京，形成了南京独特的交融的文化。

Nanjing is located at a junction of Jiangnan and Jiangbei with Yangtze River flowing in the front and leaning against the top of the Zijin Mountain. Nanjing gains majesty of the mountain and elegance of the water. It suffers from ravages of war for many times and re-arises from the rubble every time. The history of Nanjing is like ups and downs of human lifetime, and it includes both highs and lows. It was once the Jinling of flourishing age of Southern Tang and had experienced disasters brought by massacre. All of these brings up the current Nanjing and forms the unique blended culture of Nanjing.

　　一个有趣的发现，制造兵器的兵工厂却以佛教中象征吉祥海云瑞相的卍字符作为自己的标志。在回溯这块土地的历史时，我们发现了一个有趣的轮回，在曾经的西天寺遗址上兴建的金陵机器制造局，到改革开放后承担着铸造大型佛像的工作。兵工文化与佛教文化本是杀与止杀的矛盾，以及战争与和平的对立，却在这片土地上微妙的交融在了一起。

There is an interesting find that the arsenal which manufactures arms takes a symbol 卍 which represents auspiciousness in Buddhism. By looking back upon the history of this piece of land, we find an interesting reincarnation: Jinling Manufacturing Bureau built on the relics of ever Xitian temple started to cast large-scale figure of Buddha after the reform and opening up. The ordnance culture and Buddhism culture originally have the relation of killing and stopping killing, and the relation of war and peace; but, they are blended with each other delicately on this land.

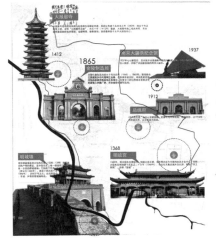

084

Industrial Heritage Protection and Interior Design　工业遗产保护与室内设计

方案设计
Scheme Design

结构分析
Structure Analysis

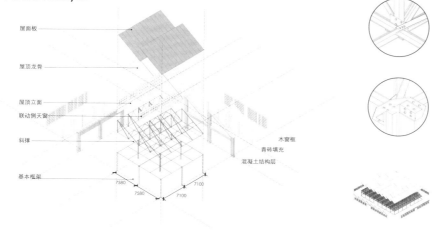

承重结构与维护结构为独立体系。重复的7.1X7.5为柱网的轻钢结构单元，通过复制形成了大空间，所有的构建以螺栓连结，易拆装，目前两栋建筑结构及外立面保存情况良好。

Load-bearing structure and maintenance structure are independent systems. The reduplicative 7.1X7.5 is a lightweight steel construction unit of a column grid and a big space is formed by duplication. All of the construction is linked by bolts and thereby can be easily disassembled. The preservation situation of structure and outer facade of the present two buildings are excellent.

整体设计
Overall Design

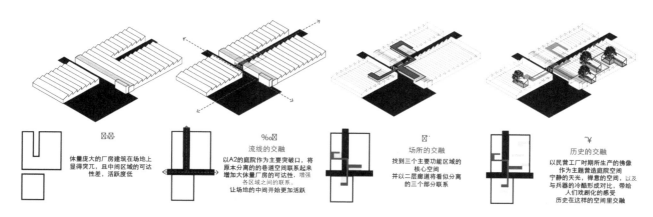

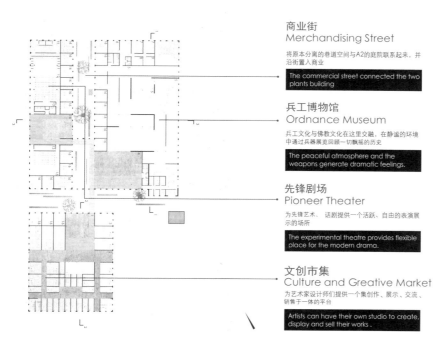

商业街
Merchandising Street

将原本分离的巷道空间与A2的庭院联系起来，并沿街置入商业

The commercial street connected the two plants building

兵工博物馆
Ordnance Museum

兵工文化与佛教文化在这里交融，在静谧的环境中通过兵器展览回顾一切飘摇的历史

The peaceful atmosphere and the weapons generate dramatic feelings.

先锋剧场
Pioneer Theater

为先锋艺术、话剧提供一个活跃、自由的表演展示的场所

The experimental theatre provides flexible place for the modern drama.

文创市集
Culture and Creative Market

为艺术家设计师们提供一个集创作、展示、交流、销售为一体的平台

Artists can have their own studio to create, display and sell their works.

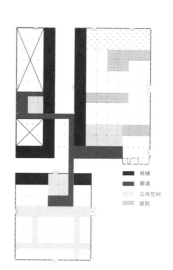

085

具体设计
Specific Design

商业街
Merchandising Street

体量庞大的厂房建筑在场地上显得突兀，且中间区域的可达性差，活跃度低。通过将连接部分改为半室外空间，将原本分离的巷道联系成为一个整体的系统，沿街置入商业，让原本沉寂的街道重新变得活跃。

The factory buildings of large size seem obtrusive on the field, having poor accessibility to the central region and a low degree of activity. By changing the connecting part to a semi-outer space, the originally separated galleries is connected to an overall system including businesses arranged down the street, so that the originally quiet streets re-become alive.

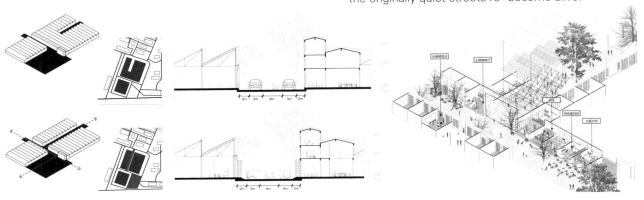

兵工博物馆
Ordnance Museum

通过置入庭院，让原本巨大的工厂空间被活化，同时作为组织流线的重要元素，在庭院的处理上，我们将屋面部分移除，并将移除的构建置于庭院中，各个庭院均以晨光机器厂时期的佛像作为营造的主题，与展览厅部分形成了戏剧性的对比，增强了人们的观展体验，对于战争的厌恶与对和平的向往。

By placing the courtyard therein, the originally huge factory space is activated and at the same time, serves as an important element of organization streamline. On processing the courtyard, we partially remove the roof surface and place the removed construction in the courtyard. Each of the courtyards takes the figure of Buddha at the period of Chenguang machine factory as the theme for constructing, which forms a dramatic comparison with the exhibition hall part, enhances people's exhibition-viewing experience, disgust to the war and looking forward to peace.

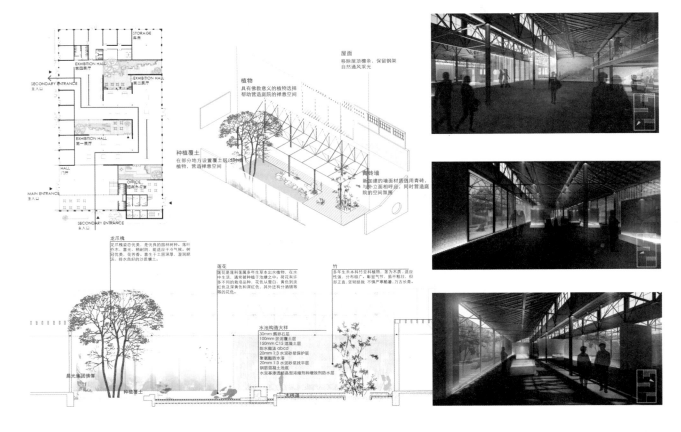

方案设计
Scheme Design
先锋剧场
Pioneer Theater

先锋剧场是当代一种新式的剧场形式，一般实验剧场面积较小，方便演员与观众的交流体验，专业的演出设备与配套设施可充分满足各种小型艺术演出与团体活动的要求；它上演的内容体现了这个时代较为先锋前卫主流的意识形态。

Pioneer theater is a new-style theater form in modern time. In general, experimental theater has a small area so actors and actresses can communicate with the audience conveniently. The professional performance equipments and supporting facilities can sufficiently satisfy requirements of different small-scale art performances and group activities. The content to be staged embodies a pioneer avant-courier mainstream ideology of this time.

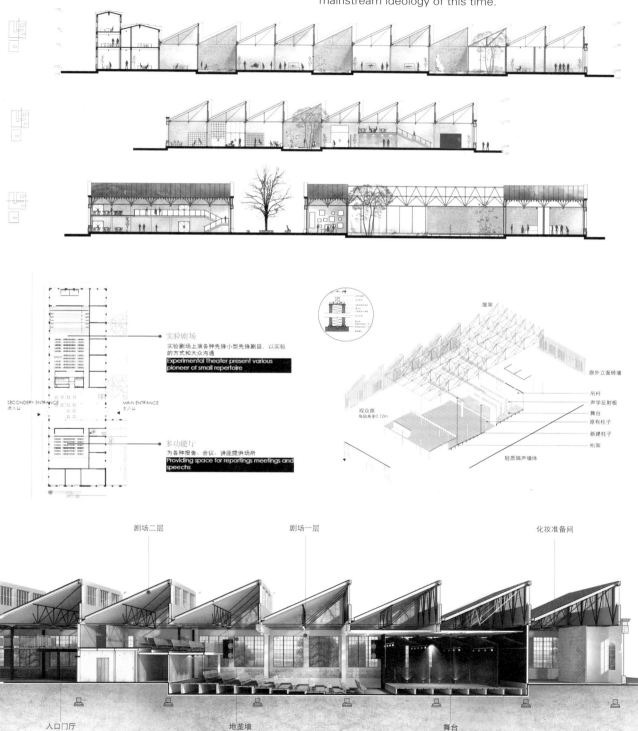

具体设计
Specific Design

文创市集
Culture and Creative Market

文创市集突破了传统的创意街区单一的功能和平面布局，在市集内建立完整的公共空间体系，满足了文创工作者设计，展示，销售，交流的多种功能。

Culture and creative market breaks through the single function and plane layout of the traditional creative park. An integrated public space system is established in the market to satisfy multiple functions of designing, showing, selling and communicating of the culture and creative workers.

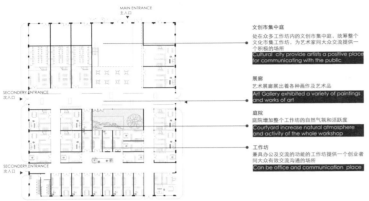

工作坊
Workshop

根据不同的使用功能及空间形态，同时由建筑本身柱网所决定，我们研究了三种不同尺度的工作坊，配合其定义加以设计。

According to different use functions and spatial forms and depending on the column grid of the building itself, we studied three kinds of workshops of different scales and designed based on the definition thereof.

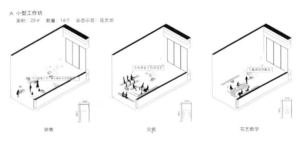

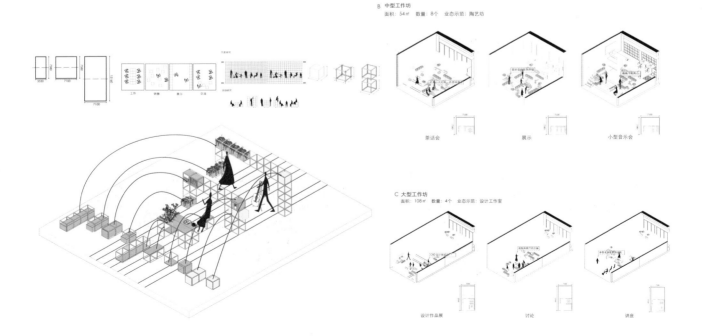

学生感想
Students' Thoughts

通过参加这个联合毕业设计，做一个自己从未接触过的题目，让我从建筑设计到室内设计学到了很多东西。学会用建筑的思维去思考室内设计，并从室内设计的角度丰富自己的建筑空间。在设计过程中了解了南京这座古老的城市，学习了旧工业厂房改造与历史建筑保护方面的知识。而且在与不同院校的交流合作中，了解了不同院校的专业特点，有了向其他院校的同学和老师学习的机会。感谢老师的指导和CIID协会创造的机会，希望"6+1"联合毕设可以越办越好，给不同院校之间创造更丰富的学习交流的机会。

According to the joint graduate project, I did a project that had never gotten in touch before. It taught me a lot from the Architectural Design to the Interior Design. I learned to use the thought of the building to think about the Interior Design, and enrich my architecture space in the view of the Interior Design. When doing the project, I understood Nan Jing, a city with a long history. What's more, I got some knowledge about old factory's reconstruction and historical building protection. According to the cooperation with different universities or colleges, I understood their characters and acquired an opportunity to learn from their teachers and students. I express thanks to teachers for their guidance. And thanks the CIID Association for giving me the opportunity. I hope the 6+1 Joint Graduate Project will be better, and give more opportunities for different universities or colleges to study and communicate with each other.

The group selects "integration" as the topic. After doing research on the history of the 1865 Theme Park and investigating on site, the two lines: war and peace, run through the environment space design of new business form, and focus on the business street, ordnance museum, Xian Feng theatre and culture creation market to develop design. The group of the concepts is completed; the expression of the design and drawing is clear and fluent. However, it's short of analysis and explanation of early business. As the result, it lacks the evidence of setting, layout, and the area distribution of the new business form in the report. Besides, it remains open to the question to remove the current houses to build several scattered small courtyards to organize the idea in the ordnance museum area. One reason is that the view wall of the court would reduce the number of the wall of exhibition hall in the museum. What's more, the light and shade from the interior and outdoor natural sunshine and manual lighting would influence the museum exhibition effect. It should be cautious to use it.

——Zuo Yan

专家点评
Experts' Comments

以"街、廊、庭"分别实现"流线的交融、场所的交融、历史的交融"，显示出设计团队对于项目的时间投入与扎实的学习能力。院落与顶部采光的变化，使各使用空间具备实现交流的自由度。构想大胆，又能相对落于实处，表达完整。

——陈卫新

According to the "street, porch, courtyard", the project team tried to make the "streamline integration, site integration, and history integration" come true. It shows their time spending to the project and strong learning abilities. The lighting in the courtyard and the top light changed. It makes the each of the using space be free to communicate with each other. The thought is adventurous, and it could be relatively put into practice, and fully expressed.

——Chen Weixin

教师点评
Teachers' Comments

本组以"交融"为题，对南京1865园区的历史和现场调研解读后将战争与和平这对立的两条线贯穿于园区新业态功能的空间环境设计，并重点对商业街、兵工博物馆、先锋剧场和文创市集四个区域展开深入设计。该组整体概念较完整，设计阐述和图纸表达清晰流畅，稍显不足的是缺乏前期业态分析说明，致使最后汇报成果中对于新业态功能的设置、布局及面积配比缺乏依据。此外，兵工博物馆通过移除现有屋面打造的多个分散小庭院来组织动线的想法有待商榷，一方面庭院观景墙面会使博物馆展厅墙面减少且不连贯，另一方面，室内外渗透带来的自然光线和人工照明的明暗交错会影响博物馆的展陈效果，须谨慎使用。

——左琰

工业遗产保护与室内设计 Industrial Heritage Protection and Interior Design

CIID "室内设计 6+1" 2015（第三届）校企联合毕业设计
CIID "Interior Design 6+1" 2015(Third Scssion)University and Enterprise Joint in Graduation Design

三等奖
Third Prize

止戈为林
Stops Fighting and Plant Trees

高　　校：哈尔滨工业大学
College：Harbin Institute of Technology
学　　生：乔红　韩思宇　贾思修　王岩　王子轩
Students：Qiao Hong　Han Siyu　Jia Sixiu　Wang Yan　Wang Zixuan
指导教师：马辉　刘杰
Instructors：Ma hui　Liu Jie

关联页：90-99、162-169

Industrial Heritage Protection and Interior Design　　工业遗产保护与室内设计

乔红 Qiao Hong　　韩思宇 Han Siyu　　贾思修 Jia Sixiu　　王岩 Wang Yan　　王子轩 Wang Zixuan

关系阐述
Relationship

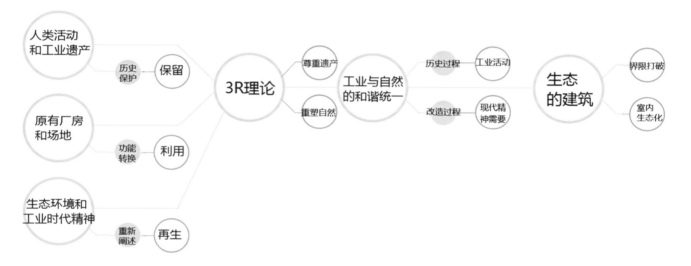

工业遗产的精神与生态的再生

历史空间向现代自然空间的过渡
——在色彩上、功能表达上的鲜明的对比，
在功能上再加以联系
暗与亮——针对人群心理环境的设计

Spirit of industrial heritage and ecological regeneration

The transition of historical space to modern natural space
——In color, a sharp contrast to the function,
In function to contact,
Dark and Bright——Design of psychological environment for people

A2　　A1

工业遗产保护与室内设计　Industrial Heritage Protection and Interior Design

方案生成
Scheme Generation

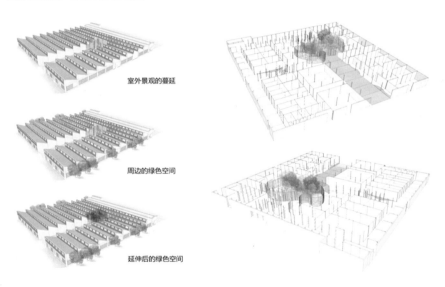

室外景观的蔓延

周边的绿色空间

延伸后的绿色空间

方案分析——照明分析
Scheme Analysis——Illumination Analysis

自然采光

——通过对屋顶的改造引入更多的自然光线；在入口处、互动区等主要区域引入自然光进行照明。

——主要照明形式：射灯、吊灯等灯具的结合使用。

Natural lighting

——Through the introduction of more natural light into the entrance of the roof of the transformation of natural light in the main area of the introduction of natural light artificial lighting.

——The main form of lighting: using a combination of spotlights, chandeliers and other lighting.

Industrial Heritage Protection and Interior Design　　工业遗产保护与室内设计

Reviving in the Dusk to Dawn

止戈为**林**

工业遗产下的精神与生态的再生

中心景观带分析
Central Landscape Zone Analysis

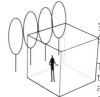

室外景观区与室内人物的关系——一定距离感

The relationship between the outdoor landscape area and the indoor character--- a sense of distance.

室外人物观室内景——减少距离感

People outside and interior view scene—to reduce the sense of distance.

植物形成某一景观区域——亲近自然之人，可选择与自然接触少距离感

Forming a certain landscape area by plants—the people who love nature can choose to contact with the nature

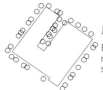

周边的种植的保留与再生

Reservation and regeneration of the surrounding plants.

建筑结合中庭形成景观区，场地的绿色向室内延伸

Combining the building with the courtyard to form the landscape zone, wherein the green of the field extends into the interior.

绿色空间对场地内有影响，植物会对室内温度和湿度产生一定影响

The green space has influence on the field and the plant will generate a certain influence on indoor temperature and humidity.

原有室内空间

The original indoor space.

引入绿色植物，模糊室内外界限

Introduce the green plants to blur a boundary between outdoor and indoor.

避免室外的影响——隔断

Avoiding influence from the outdoor—separation.

柱网分析
Column Grid Analysis

室内原有柱网结构

Interior original column grid structure

不做减法做加法，通过柱网的组合引导走向

Rather adding than reducing to guide orientation by combining the column grids.

形式一
引导人群，通向不同的区域

Form I
Guiding crowd to different areas

模型展示
Model Exhibition

植物形成某一景观区域——不亲近自然之人，可选择不予理睬

A certain landscape area formed by plant---the people who doesn't like nature can choose to neglect it.

将形式结合到室内空间，在室内感受到室外生态的延伸少距离感

To combine form with the interior space so that people can feel extension of the outside ecology in the interior space.

与室内结合形成半开放空间，增强室内外的交流

Indoor combination of semi open space, enhance the exchange of indoor and outdoor.

将中心景观区与室内进行一定的隔绝，形成中心玻璃体隔断，减小对室内的影响

Make certain separation between the center landscape zone and the interior space to form a center glass partition so as to reduce the influence for interior space.

折线形的玻璃体给人尖锐的感受，弧线形与刚硬的工业厂房结构形成对比并对屋顶进行一定的改造

The glass body in the shape of folded cables gives a sharp feel to people. The shape of arc contrasts with the rigid industrial factory.

将弧线进行一定的组合形成最终的形态，并留有三个入口

Combining the arc lines to form the final shape and leaving three entrances.

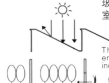
玻璃隔断有入口，形成室内外过渡区域

The glass partition has an entrance to form an outdoor-indoor-transitional area.

对屋顶改造，使阳光更充足，植被不受限制

On the roof of the transformation, so that the sun is more abundant, the vegetation is not limited.

最终形成连接室内外的景观过渡空间

Finally, the indoor-outdoor connecting landscape transition area is formed.

形式二
结合植物分布，在柱网上缠绕植被，形成多处绿色空间

Form II
Combining with plant distribution, the column grid is wound by vegetation to form green spaces at many places.

形式三
高矮不一的柱网，融入不同的功能，与灯具结合

Form III
The column grids having different heights are added with different functions to combine with lamps.

形式四
在行走路线中加入展示牌与A1结合作为历史介绍

Form IV
Adding display cards in the walking line to combine with A1 as introduction of history.

内墙改造
Reconstruction of the Interior Wall

内嵌改造区
Embedded reconstruction area

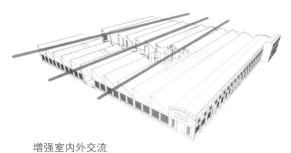

增强室内外交流
模糊界限
To enhance indoor-outdoor communication blurring limit

消除空间的孤立感
带动发展
Eliminate the space of the isolated sense to drive development

屋顶改造
Reconstruction Area of the Lnterior wall

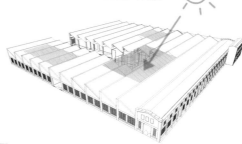

屋顶改造区
Roof reconstruction area

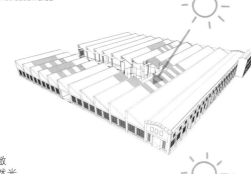

部分开敞
引入自然光
Be partially open to introduce natural light

部分改造
植被可生长
The partially reconstructed vegetation can grow

平面图
Plane Graph

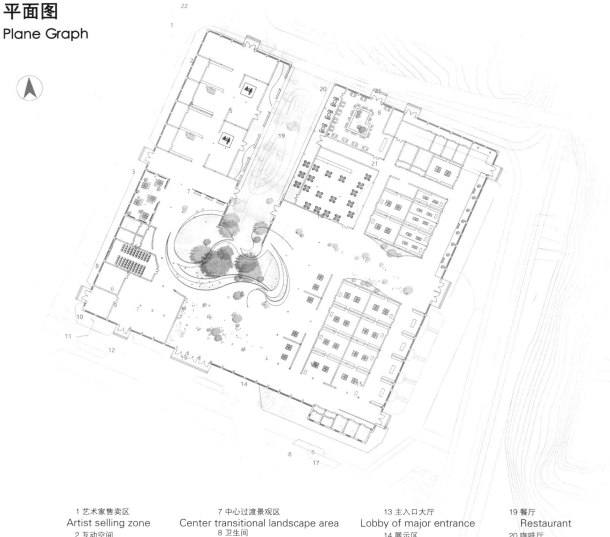

1 艺术家售卖区 Artist selling zone	7 中心过渡景观区 Center transitional landscape area	13 主入口大厅 Lobby of major entrance	19 餐厅 Restaurant
2 互动空间 Interactive space	8 卫生间 Rest room	14 展示区 Exhibition area	20 咖啡厅 Coffee house
3 艺术家私人空间 Private space of artists	9 小型放映厅 Small-scale video hall	15 对外售卖处 Outward selling area	21 设备间 Equipment room
4 过渡半室外空间 Transitional semi-outdoor space	10 放映厅设备间 Equipment room of the video hall	16 商铺 Shop stores	22 广告展示区 Advertisement display area
5 室外互动空间 Outdoor interactive space	11 仓库 Warehouse	17 办公室 Office	
6 书店 Bookstore	12 展卖区 Exhibition for selling area	18 次入口大厅 Lobby of secondary entrance	

A1、A2 剖面分析
A1、A2 Profile Analysis

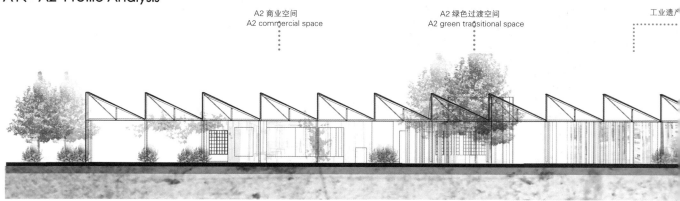

Industrial Heritage Protection and Interior Design | 工业遗产保护与室内设计

平面图分析
Plane Graph Analysis

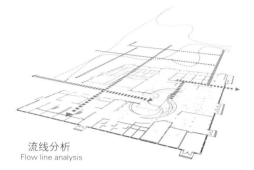

流线分析
Flow line analysis

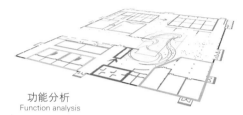

功能分析
Function analysis

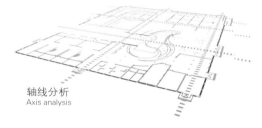

轴线分析
Axis analysis

灯位布置
Light point arrangement

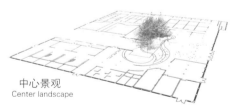

中心景观
Center landscape

庭院分析
Courtyard Analysis

儿童与植物互动
Interaction between children and plants

休闲娱乐
Recreation & Entertainment

植被缓坡
Vegetation gentle slope

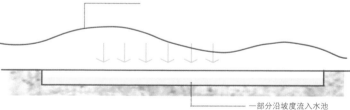

一部分沿坡度流入水池
A portion of rain water flows into a pool along the gentle slope

——历史空间向现代的自然空间的过渡—— ——形成对比、呼应

A1 历史博物馆
museum

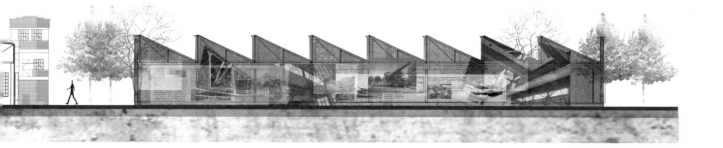

097

中心景观区域
Center Landscape Area

冰冷的钢铁和砖石中增添了些许绿色，使枯燥的厂房变得有趣、清新。外与内的界限不再清晰，生态精神在工业遗产中再生，在刚与柔的空间中带给人以独特的感受。

The cold steel and masonry are added with green so that the boring factory buildings become interesting and fresh. The limit of the outdoor and interior spaces is no longer clear. The ecological spirit is regenerated in the industry heritage to bring people special feeling in the hard and soft space.

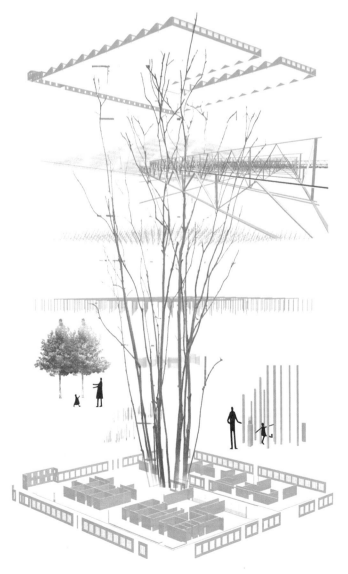

学生感想
Students' Thoughts

很荣幸可以参加CIID"室内设计6+1"2015（第三届）校企联合毕业设计活动，有机会可以和其他六所知名的高校进行切磋学习，得到专家对我们自己的设计的点评和建议。这是一个很难得的机会。通过开题、中期以及终期各个学校在一起的交流，每个学校各具特色的设计拓宽了我们的思路，也开阔了我们的视野，伴我们一路成长。时间虽短暂，这一路走来，我们并不是只有苦恼和疲劳，也拥有了很多欢乐的时光。所以，我们享受了这个过程。

It's my honor to take part in the CIID "Interior Design 6+1" 2015 College & Enterprise Joint Graduate Project. I had the chance to communicate with other 6 colleges or universities. It's also a rare opportunity to get comments and suggestions from the specialists about our own projects. According to the communication with each other among these colleges during the opening speech, mid-term examination and final reply, our thoughts were expended, and our visions were broadened. The project followed our growth. Although the time was short, we did not only have trouble and tiredness, but also own happiness. So we enjoyed ourselves during the period.

教师点评
Teachers' Comments

"止戈为林"的主题，很贴切1865工业厂房改造项目，即显示了1865厂房过去的历史、文化属性，又明确了改造方案的主题发展方向——通过室内植物"林"的营造，来突出平和、绿色的展示品性。方案的构思主题是明确的，在手法上也有所表达，遗憾的是在"林"的空间营造中，显得太过平淡，主题在各个空间中没有得到很好的强调和显示，如能把各个展示空间的改造与"林"的构思结合得更密切，会使方案的逻辑关系更强。

方案的表达比较完善，包括建筑改造、室内空间设计、景观设计的内容都很充实，工作盈满。值得肯定的是对建筑技术的应用有一定的思考。

——陈建华

The subject "cease harm towards the forest" get close to the meaning of the 1865 Industrial Factory transform project. It shows the history and cultural property of the factory, and ensures the theme of the transform project--- constructing the forest of the plants indoors to highlight peace and green as its exhibition property. The thought of the theme is clear, and the technique is expressive. However, it's too insipid during the creation of the forest space. The theme is not showed and emphasized well in all the spaces. If the expression of the transform in each exhibition space could combine the forest thinking more closely, the project will be more logical.

The expression of the program is completed. The content is full, including the construction transform, the inner space design, and the landscape design. It is sure that it is indicative in the construction field.

——Chen Jianhua

专家点评
Experts' Comments

"止戈为林"，从精神层面和自然层面，结合历史和现状，拓展了对"金陵兵工"的理解。形而上与形而下的结合，两种维度的解读，将A1和A2为之关联，让人眼前一亮。

对近代兵工博物馆的设计和A2业态布局、空间构成、景观布置的生成，逻辑思路清晰，考虑周详；对锯齿形厂房特殊的建筑肌理剖析到位，柱子加密的逆向思维让整个方案增色许多。整个方案让建筑内外的环境诗意模糊，不断追寻历史精神的同时，又能穿梭于生态的空间，乃是概念上乘、设计上乘的好作品。

——邹超

Combined with history and current situation, "cease harm towards the forest" expands the understanding of the "Jinling Ordnance" in the spiritual and natural levels. The combination of Metaphysics and Physics, and the understanding of the two dimensions, make a connection between A1 and A2. It makes an impression on the visitors.

It's logical and considerate to create the design of the modern ordnance museum, the A2's business layout, the space composition and the landscape layout. The analysis of the zigzag factory's special context is accurate. It's a reserve thinking to increase the density of pillars. That increases the attraction of the project. The whole project makes the environments inside and outside the building become poetic. It does not only seek for the historical spirit, but also shuttle back and forth in the ecological space. It's a good work in the concept and design.

——Zou Chao

工业遗产保护与室内设计 Industrial Heritage Protection and Interior Design

CIID "室内设计6+1" 2015（第三届）校企联合毕业设计
CIID "Interior Design 6+1" 2015(Third Scssion)University and Enterprise Joint in Graduation Design

三等奖
Third Prize

生长和碰撞
Growth and Crash

高　　校：	哈尔滨工业大学
College:	Harbin Institute of Technology
学　　生：	金喆　伏祥　何佳佳　张泽宇
Students:	Jin Zhe　Fu Xiang　He Jiajia　Zhang Zeyu
指导教师：	马辉　刘杰　周立君
Instructors:	Ma hui　Liu Jie　Zhou Lijun

关联页：100-105、169-173

金喆 Jin Zhe　　伏祥 Fu Xiang　　何佳佳 He Jiajia　　张泽宇 Zhang Zeyu

碰撞与生长——南京 1865 产业园 A2 室内设计
Growth and Crash——Industrial Park A2 Interior Design of Nanjing in1865

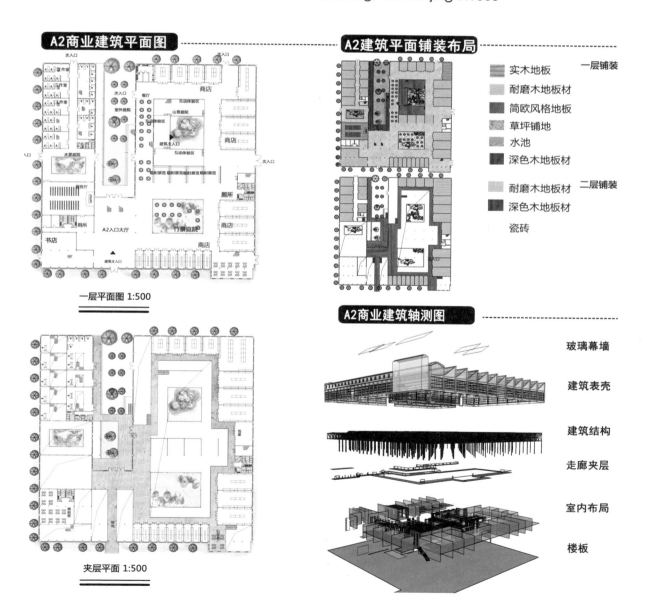

工业遗产保护与室内设计 Industrial Heritage Protection and Interior Design

碰撞与生长——南京1865产业园A2室内设计
Growth and Crash——Industrial Park A2 Interior Design of Nanjing in1865

室内色彩规划意向

在本次设计中，我们采取了原有的建筑风格元素，通过改造，融入了透明的玻璃等和原有建筑不相同的元素，试图营造一个崭新的总和商业空间。

In the design, we adopt the original architectural style elements and add the transparent glass and other elements that are different from the original architectures by reconstruction to try to create a new overall commercial space.

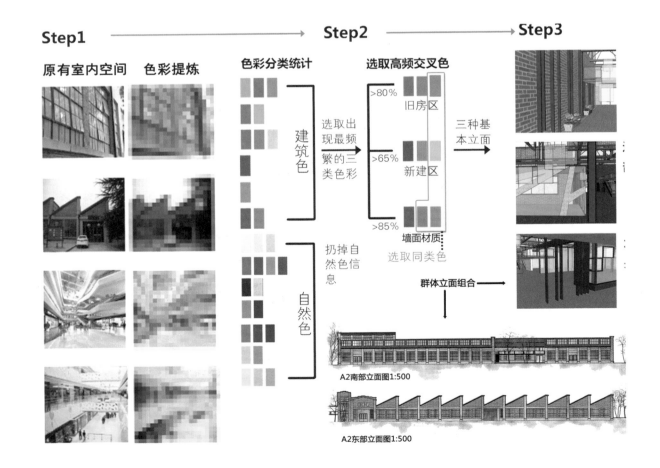

碰撞与生长——南京 1865 产业园 A2 室内设计
Growth and Crash——Industrial Park A2 Interior Design of Nanjing in1865

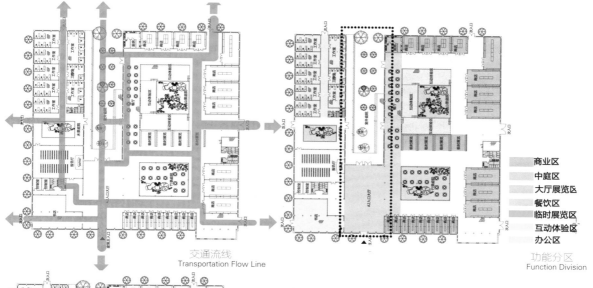

交通流线
Transportation Flow Line

功能分区
Function Division

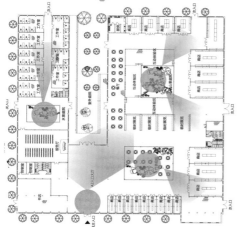

视线分析
Line of Sight Analysis

通过分析工业改建成商业综合体的设计要素，提炼了以三个中庭为主要结构中心体系的设计思路，结合建筑本身环境，生长于碰撞的概念生成，整个设计夹层的走廊贯穿始终，巧妙的增加垂直交通空间变化，使三中庭与结构体系相结合，打造独特室内效果。

By analyzing main design elements of reconstructing industry to commercial complex, we determine a design idea which have three courtyards as the main structure center system; And combining with the architecture environment and concept generating of growth and crash, the corridor of the entire design interlayer goes throughout. The vertical transportation space change is increased smartly to make three courtyards combine with the structure body so as to create unique interior effect.

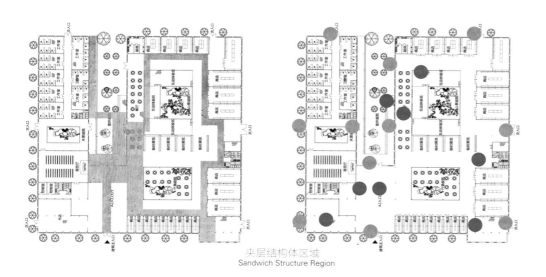

夹层结构体区域
Sandwich Structure Region

碰撞与生长——南京1865产业园A2室内设计
Growth and Crash——Industrial Park A2 Interior Design of Nanjing in1865

1 竹 Bamboo

竹中庭位于整个室内核心区域，将江南的竹融入到整个中庭当中，将采光与植物相结合，减少工业带来的气息。

Bamboo atrium is located in the core area of the whole interior, will be integrated into the entire atrium of the south of the Yangtze River, and the combination of light and plant, reducing the industry brings breath.

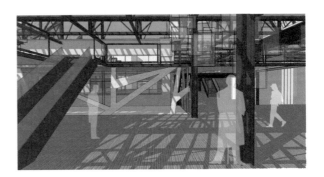

2 水 Water

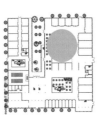

3 山 Mountain

学生感想
Students' Thoughts

这次的设计是由我们的艺术设计系的学生和建筑系的学生共同完成的。大家毕业后的去向不一样，各人设计的思路也不一样，但是在四个完全不同的大脑的碰撞之后，我想我们交出了我们满意的答卷。

我们组的四个同学，三个来自江苏，其中伏祥同学就来自于南京，我们对于南京都非常的熟悉，这给我们提供了设计的激情和动力，我们都有改造这一城市、这一场所的设想。在这个具有历史意义的建筑中，我们组采取了寻找场所的历史来由，对场所的文脉进行抽丝剥茧的分析，从而提取场所独一无二的特点的做法，利用加法、减法共同作业的方式，完成了本次的设计。

我想，我们这次的设计中最让我们满意的就是设计的一体性。我们每个人都没有各自为战，而是将相互的设想结合在一起，我想，这对于我们来说，都是一次令人兴奋的旅程。

This project was altogether finished by the students from our Art Design Department. After graduation, their purposes are different, and their thoughts are different, too. However, I think we get a good result for us through the four people's efforts.

Our group contains four members. Three of us are from Jiangsu. Fu Xiang, a member of ours, is from Nanjing. We are familiar with Nanjing. It gives us the motivation and passion. All of us have thoughts about the city's reconstruction, and the site reconstruction. In the building with historical significance, our group sought for the historical source and analyzed the context in detail. Then we got the unique feature of the site. We finished the project through addition & subtraction joint operation.

I think, the point we satisfy most is the unity of the thought in the design period. Neither of us worked alone. Instead, each thought was combined with each other. I think, it's an exciting journey for all of us.

部分，面积较大，基地作为遗产保护的建筑，如果能将其保留、归纳收藏并进行展示，该设计方案会更加完整、更具说服力。

——何方瑶

The author of the graduated work Growing and Collision understood the history context thoroughly, and analyzed it in detail. The content is abundant, and its expression is sufficient and strong. The originality is fresh and completely expressed.

When designing the A1 building, the author used the historical development of the Buddhism culture around the Nanjing Chen Guang Armory as the clue, and meanwhile explored the development history of the Chen Guang Armory. He followed the factory's history to carry out the display design. The A1's function layout is clear; its showing streamline is fluent. Part of the building is designed two floors. It spreads the showing room. The showing method is abundant, and the showing result of each room is fine. Besides, the work also mentions the light design of exhibition hall. However, there exists some shortage. The theme of the exhibition hall involves in both Buddhism culture and Chen Guang Armory, and the streamline refers to the history of the Buddhism and Chen Guang Armory. The two factors' relation is thoughtless, and the showing thought is confused. It's too abrupt to bring the Buddhism culture in.

Based on the remained construction, the A2's design keep the ancient trees on site, remove parts of the buildings' roofs and replace the glass material. It creates some courtyard space, and transits half of the outside space. A2's function division is clear; the traffic streamline is mainly annular interchange, connecting first and second floors and their function areas. It's clear and fluent. The stone, bamboo, and water, three themes of the space are fresh and creative. They also involve in inner lighting, color, material design, and joint structure of the courtyard space. The design is deep. However, the shortage is that the effect of the A2 space express is not good. It is not fully expressed. Meanwhile, the area of the removed roofs is large. As the protective heritage, if the base could be reversed, collected and exhibited, the project would be more integrated and convictive.

——He Fangyao

教师点评
Teachers' Comments

毕业设计作品《生长＆碰撞》对基地历史文脉挖掘较为透彻，并做了详细分析。设计内容丰富，表现较为充分，有力，设计创意新颖，表达完整。

A1设计中，以南京晨光厂址旁佛教文化的历史发展为线索，同时挖掘晨光厂的发展历程，以厂史发展为主线进行展示设计。A1功能布局清晰，展示流线顺畅，局部做二层设计，拓宽了展示空间。展示手段较为丰富，对于具体各空间的展示效果有不错的设计表现，对展厅的照明设计使用也有提及。不足的是，展馆的展示主题同时引入佛教文化与晨光厂文化，流线以佛教文化历史与晨光兵工厂历史交织进行，对于两者之间的关系考虑欠妥，展示思路有些混乱，佛教文化的引入过于突兀。

A2设计中，在保留建筑结构的基础上，保留场地古树，拆除部分建筑屋顶以玻璃材质代之，创造出一些庭院空间，以及过渡半室外空间。A2功能分区明确、交通流线以环形立交为主连接一二层及各功能区，清晰流畅。石、竹、水三个主题空间构思新颖，有一定设计创意。也涉及了室内照明、色彩、材质设计，庭院空间的衔接结构设计，设计有一定深度。不足的是，A2空间效果表现欠佳，空间的效果表达不够充分。同时，庭院部分拆除的原建筑屋顶

专家点评
Experts' Comments

方案对原有厂房的实际情况做了充分的分析，根据改造的功能要求和人的活动需要进行了有序的空间设计。设计逻辑关系明确，功能布局合理，流线组织清晰，空间层次分明。设计通过增减的手法，使原来的厂房增加了丰富有序的空间，增减部分的结构处理和材质运用能充分顾及原有厂房的风貌。室内外空间增设了很多绿化庭院，令各功能区域更为宜人。室内展陈空间设计上手法较为单一，考虑略有欠缺，在室内展陈氛围的营造上存在提升的空间。

——冼剑雄

The project fully analyzed the current situation of the original factory. According to the requirements of the reconstructive functions and people's activities, the space design was well-organized. The logical relation of the project is clear, the layout is reasonable, the streamline is clearly organized, and the space level is clear. The original factory added some ordered and abundant rooms according to the increase and decrease method. The process of the increased or decreased parts and the usage of material could fully take the previous scene of the factory into account. The project increased many green courtyards inside and outside the space. It makes each of the function areas be pleasant. However, the design method of the exhibition inside is too simple. The thought is not completely considerable. The construction of the exhibition atmosphere inside should be improved.

——Xian Jianxiong

工业遗产保护与室内设计 Industrial Heritage Protection and Interior Design

CIID"室内设计 6+1" 2015(第三届)校企联合毕业设计
CIID "Interior Design 6+1" 2015(Third Scssion)University and Enterprise Joint in Graduation Design

三等奖
Third Prize

渠水流方
Introducing the Flowing Water to Interior Design

高　　校： 西安建筑科技大学
College: Xian University of Architecture and Technology
学　　生： 刘璧凝　孙宗藜　李肖路　潘岩
Students: Liu Bining　Sun Zongli　Li Xiaolu　Pan Yan
指导教师： 刘晓军　何芳瑶
Instructors: Liu Xiaojun　He Fangyao

关联页：106-111、203-207

| 刘璧凝 | 孙宗藜 | 李肖路 | 潘岩 |
| Liu Bining | Sun Zongli | Li Xiaolu | Pan Yan |

设计说明
Designing Explanation

把现代人的价值观念、生活方式以及对环境的需求作为改建的基本要求 "回归自然"——通风、采光、环境以简洁清晰的设计手法去表现现代社会的生活价值观朴素、简洁的美学观念用易于理解的甚至是可逆的处理方法表达对历史的尊重。保存场所精神，体现历史文脉以保留为主，改造为辅。回归自然是现代价值观和生活方式的表现。好的通风、采光和自然环境是最基本的要求。适宜的设计赋予老建筑全新的意义和明晰空间、光、气流和视觉关系。通透的体量为室内创造光线充足、视野开阔的空间氛围。

Taking people's view of value, life-style and demands for the surroundings as basic requirements for rebuilding, "returning to nature"---ventilating, lighting and the surroundings are designed in a concise and clear manner to express life values of modern society, the simple and concise aesthetic conception is processed by a easily-understood and even reversible method to express respect to the history and preserve the spirit of place, which takes historical context reservation as major and rebuilding as supplement. Returning to nature is the expression of modern values and life-style. Good ventilation, lighting and natural environment are basic demands. Suitable design will impart clear space, light, airflow and visual relations of new meaning to old buildings. The transparent mass creates a space atmosphere with enough light ray and wide view for the interior.

A2 公共空间
A2 Public Space

采用渠水的元素在地面上设计了细窄的曲形水渠。用这种方法来限定空间，平时可以用来放户外休闲座椅，也可以充当临时的展览空间。依据地上的限定空间来展放展板，人可以跨过水渠，也可以依据它的形状而行，来游览整个中庭。

Adopting an element of canal water to design a curved canal having a finely narrow width makes a fine method to define the space. Usually, it can be used to place the outdoor leisure chairs and also serves as the temporary exhibition space. Panels are placed in accordance with defining space on the ground. People can cross over the canal or walk along its shape to visit the whole courtyard.

工业遗产保护与室内设计　Industrial Heritage Protection and Interior Design

中庭设计
Courtyard Design

将云锦的元素置入到公共空间当中围和空间，用未完成状态的云锦丝线的形态来构筑具有南京韵味的空间。靠近中庭的露天餐厅上，有水流流下跌落在水池中，在一层空间入口处形成一个独特的水帘，人在行走的过程中，可以自由的穿过水帘，形成一种特别的空间感受。

Placing an element of brocade to the public space to enclose the space and constructing the space with the feeling of Nanjing using the form of brocade silk thread of unfinished tense. In the open-air restaurant close to the courtyard, water flowing down and falling off to water tank and thereby forms a unique water curtain at an entrance of the ground floor. By walking therein, people can freely walk through the water curtain to have a special feeling of space.

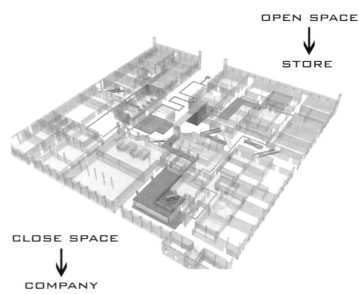

OPEN SPACE ↓ STORE

CLOSE SPACE ↓ COMPANY

A2 平面图
A2 Plain Graph

一层平面图

二层平面图

108

Industrial Heritage Protection and Interior Design　工业遗产保护与室内设计

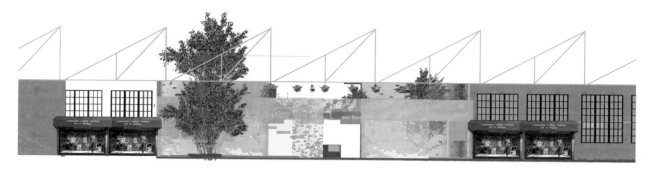

中庭立面 Courtyard Facade

公司立面 Company Facade

办公空间设计
Office Space Design

提取盒子元素，用拉丝不锈钢板来把工作空间限定出来，利用几何体巧妙地分隔空间形成独立的办公空间。同时朝外大面积开窗，自然景观尽收眼底，给人一个舒适的办公环境。

Extracting box elements and defining the work space with wiredrawing stainless steel plate. The space is skillfully separating by utilizing geometry to form an independent office space. At the same time, we design large areas of windows towards the outdoor space to have a panoramic view of natural landscape so that people can enjoy a comfortable working environment.

在整体的立方空间中套入小的立方体，与大环境呼应同时有形成了独特的办公空间，拉丝不锈钢板为员工限定了一个较为私密的空间。

Inserts a small cube into the integral cubic space to echo with the broad environment and at the same time forming the particular office space. The wiredrawing stainless steel plate acts to define a private space for the staff.

手工坊设计
Workshops Design

在手工坊中用云锦元素把每一个手工桌围和成一个个独立的空间。

Each of the handcraft table is enclosed to be an independent space using the brocade element in the workshops.

餐厅设计
Restaurant Design

通常建筑用的钢筋被转换为室内的分隔材料。钢架通常隐藏于建筑结构墙体中不被人所见，然而在重新涂成黑色并交叉搭建成网格系统后在鸿咖啡里却成为室内的焦点，使家具结构与建筑结构融为一个整体。塑造了空间整体性同时也让有限的空间变得更加深邃。

In general, rebar used for buildings are converted to the interior separating materials. The steel frame is generally hidden in the wall of building structure and thereby cannot be seen by people, but becomes an interior focus point of Coffee house after being re-painted to black and intersecting with each other to form a grid system. In this way, the furniture structure and building structure are blended to a whole, which shapes integrality of space and at the same time make the limited space become deeper.

人流功能分析
Analysis of Function of the Stream of People

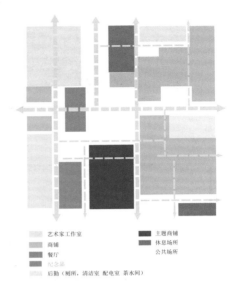

艺术家工作室　主题商铺
商铺　　　　　体息场所
餐厅　　　　　公共场所
纪念品
后勤（厕所，清洁室 配电室 茶水间）

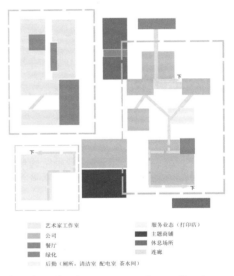

艺术家工作室　服务业态（打印店）
公司　　　　　主题商铺
餐厅　　　　　体息场所
绿化　　　　　连廊
后勤（厕所，清洁室 配电室 茶水间）

设计灵感来源于3Ds max的布尔运算，长方体的桌面经过圆柱体的裁切形成具有趣味性的办公桌。穿插一些绿植，使办公空间更加具有亲切感和创造力。

The design inspiration comes from Boolean calculation of 3Ds max. The desktop of cuboid is cut by a cylinder to form office tables with interestingness. Some green plants are placed therein such that the office space has cordial feeling and creative power.

学生感想
Students' Thoughts

经过3个月的深入，我们对这个毕业设计题目也有了一定深度的了解。城市肌理可以反映城市精神，城市肌理是在长期的浸润和积淀形成的，是一部活的城市历史，印记着岁月的沧海桑田。南京深厚的历史底蕴确实容易让人跌进去，而得以触摸其肌理当然令人迷醉。我们希望尽量去还原原本空间的特质，呈现建筑受自然侵蚀的痕迹。同时通过不同吊顶形式整合机电系统来应对开放与封闭空间效能的需求。在被修缮过的老建筑里，历史与城市变得鲜活可感，人关注自己与城市的关系，这样的情感与内心建设，是一种文化的力量。通过这次比赛，我们发现了自身许多的不足，感谢各位老师的教诲，我们会在以后的发展中越来越好。

According to the three months work, we understood the graduate project theme in a certain degree. The context of the city could reflect the city's spirit. It is formed by people's long-time activities and their habits. It represents the city's history, and tells us its stories. Nanjing's long history could really make people fall in love with it. They are amazed at its context. We hoped to recover the feature of the original space, and show natural erosive trace of the building. Meanwhile, according to different suspended ceiling patterns, we integrated the electromechanical system to meet the needs for the efficiency of the open space and closed space. In the recovered old building, the city and its history become fresh and sensible. People focus on the relations between the city and themselves. The emotion & heart construction is the cultural force. According to the match, we found many shortages of ourselves. Thanks for teachers' edification. We will become better in the future.

教师点评
Teachers' Comments

优点：
（1）．设计定位清晰：尊重历史、注重发挥场所精神、体现文脉传承、力求再生、突出文创产业园的创新。
（2）．空间构成简明：提取旧城街坊的城市肌理作为平面布局的方式，用体块为元素进行穿插扭转，分割出不同功能的空间，隐喻古城文脉，增添设计趣味。
（3）．渠水流方：从南方民居门前的水渠意向元素，形成水系设计系统，景观设计贯穿建筑内外，环境设计效果好。
缺点：
（1）．商业设计主题不鲜明：吸客力与其他创意产业园设计没有更深入和细化的突破。
（2）．平面动线设计不通达：有的店铺会形成死角。
（3）．水系体验参与性弱：处于被观赏状态，没有更多的交互体验。
（4．）展示馆辅助空间和休闲空间没有设计。
（5）．视觉传达特征不强：空间界面、环境设施、导识应协同呼应突出主题。

——杨琳

Advantage:
（1）. The position of the design is clear: Respect the history, emphasize the site spirit, express the culture inheritance, strive for the rebirth, and highlight the innovation of the cultural innovation park.
（2）. The construction of the space is simple and clear: Adopt the pattern of the old streets in the city as the way of the layout, use the block as the factor to alternate and reverse, and divide the space into several parts with different functions, describe the culture factor by metaphor and gain the interest of the design
（3）. The flow of the water channel: The image factor of the water channel in the southern dwellings outdoors forms the water design system. The landscape design run through the whole construction, the result of the environment design is good.
Shortage:
（1）. The theme of the business design is not clear: the customer attraction and other creative industry park design don't make deep and refined breakthrough..
（2）. The design of surface is not clear: Some store will from the dead angle.
（3）. The property of water system participation is weak. It's appreciated without interactive experience.
（4）. There is no assist room and rest room left.
（5）. The feature of vision transfer is not strong enough: The space interface, environment instrument, and the guidance equipment should cooperate with each other to emphasize the theme.

——Yang Lin

专家点评
Experts' Comments

方案在对所给历史建筑现场的现状、能耗、基地、区位、通风与日照、降雨、周边业态等分析的基础上进行设计，思考比较全面，特别是对光照、空气流通的思考是非常可取的。水的引入给空间增添了活力，但是对水池的安全防护和架空玻璃水道可实施性欠缺考虑，底层水景边上无提示，人在行走中容易出现安全隐患。对空间中，光照与空气条件和改变条件的数据没有充分的说明。

盒子空间组合富有想象力和视觉美感，但缺少与特定历史建筑特征的结合，对在"限定"条件下的创造力有待提高。

——余平

The project was designed based on the analysis of the current historical site, energy consume, base, location, ventilation & lighting, rainfall, surrounding business activities and so on. The thinking is considerate; especially the thoughts of lighting and ventilation are advisable. The import of water increases the vitality of the space. However, the thoughts of the water pool protection and glass water channel overhead are unconsidered. There is no warning signal near the bottom water landscape. When walking on the water channel, there exist safety risks. To the space, the data of the lighting, ventilation and changing condition is not fully explained.

The group of box space is filled with imagination and visual beauty, but it is lack of combination with features of the special historical buildings. The creativity under the limited condition should be improved.

——Yu Ping

工业遗产保护与室内设计 Industrial Heritage Protection and Interior Design

CIID "室内设计 6+1" 2015（第三届）校企联合毕业设计
CIID "Interior Design 6+1" 2015(Third Scssion)University and Enterprise Joint in Graduation Design

三等奖
Third Prize

金陵风吟
Wind Sing of Jinling

高　　校：	西安建筑科技大学
College:	Xian University of Architecture and Technology
学　　生：	冯胤云　何其畅　韩杰林　黄超　胡铭中
Students:	Feng Yinyun　He Qichang　Han Jielin　Huang Chao　Hu Mingzhong
指导教师：	刘晓军　何芳瑶
Instructors:	Liu Xiaojun　He Fangyao

关联页：112-119、208-212

Industrial Heritage Protection and Interior Design 工业遗产保护与室内设计

| 冯胤云 | 何其畅 | 韩杰林 | 黄超 |
| Feng Yinyun | He Qichang | Han Jielin | Huang Chao |

金陵风吟
Wind Sing of Jinling

区位分析
Location Analysis

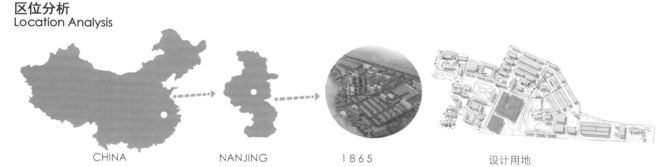

CHINA　　NANJING　　1865　　设计用地

周边建筑分析
Surrounding Buildings Analysis

1950s to 1960
pop design

1950s to 1980s
High technotogn

1960s
rationalism

1960s to 1980s
postmodernism

1980s
decons trution

1990s
gree design

2015s
modern

The future

Four-dimensional space Ganshou

南京，简称宁，江苏省省会城市。南京地处长江下游平原，国家生产力布局中最大的经济核心区——长江三角洲，宁镇丘陵山区。北纬31°14'～32°37'，东经：118°22'～119°14'，东望大海，西达荆楚，南壤皖浙，北接江淮，长江越境而过。距上海300公里，距北京1200公里，向西离重庆1400公里。

　　Nanjing, abbreviated as Ning, is a capital city of Jiangsu province. Nanjing is located at a plain downstream the Yangtze River, the largest economic core zone in the location of national productive forces——Yangtze River delta, hilly and mountainous area of Ning town. At northern latitude: 31°14'～32°37' and east longitude: 118°22'～119°14', Nanjing looks eastward across the sea and reaches westward to Jingchu; it is adjacent to Wan-Zhe at south and Jianghuai at north. The Yangtze River flows across over the Nanjing. It is 300 km away from Shanghai, 1200 km away from Beijing and 1400 km away from Chongqing westward.

金陵风吟
Wind Sing of Jinling

竖向分析
Vertical Analysis

周边建筑

主要道路

主要绿植

人群聚集

夏天风向

冬天风向

降水聚集

日照

植被

（1）."以人为本"，创造舒适宜人的可人环境，体现人为生态。"人"是景观的使用者。因此首先考虑使用者的的要求、做好总体布局。本次景观设计大量加入集散区域以及休闲广场，旨在提高公共空间的同时使人们最大可能的得到休闲放松。室内外串联的下沉广场不仅丰富了室内空间，也最大程度的扩展了人们休闲娱乐的空间。

（2）."以绿为主"，最大限度提高绿化率，体现自然生态。
设计中加入大量绿植，树木。主要采用以植物造景为主，绿地中配置高大乔木，茂密的灌木，营造出令人心旷神怡的环境。绿色不仅给人带来轻松愉悦的心情，也为南京城市绿化做出标榜。

（3）."崇尚自然"，寻求人与自然的和谐。
此次设计在有限的生活空间利用自然、师法自然，加入木栈道、水晶小品、景观凉亭、人工湖等寻求人与建筑、山水、植物之间的和谐共处，使环境有融于自然之感，达到人和自然的和谐。A1、A2建筑之间加入冥想堂，旨在创造空间的基础上，更加通风贯通，更重要的是让人们在冥想堂卸去铅华，调整状态，更认真的投入生活与工作。

1. "Residence oriented" to create a comfortable and delightful environment and embody man-made ecology. The "human" is the user of landscape. Thus, the requirement of the user must be considered at the first to make good overall layout. The landscape design at this time is added with a large number of gathering and distributing regions and leisure squares, aiming at improving utilization of the public space and make people have relaxation at the utmost. The indoor and outdoor sunken squares connected in series not only enrich the indoor space but also lift the space for entertainment of people at the greatest extent.

2. "Focus On Green" to Maximize the Greening Rate and Embody Natural Ecology. The design includes planting a large number of green plants and trees. Mainly adopting plants for landscaping, the tall and big arbors are planted in the central green land so that the thick bush creates a relaxed and happy environment for the people. Green color would bring people relaxed and cheerful mood and being a role model for urban landscape of Nanjing.

3. "Advocating the Nature" to Seek for Harmony between Human and Nature
The design of this time utilizes nature, learning from nature, joining wooden trestle, crystal sketch, landscape pavilion, artificial lake and the like in the limited life space to seek for harmonious coexistence among human, landscape and plants, so that there is a sense of environment blended into the nature and the harmony between human and nature is achieved. A meditation hall is placed among the A1 A2 buildings, aiming at ventilating and cut-through on the basis of space creation. More important, people can return to innocence and adjust the state in the meditation hall to put into life and work more earnest.

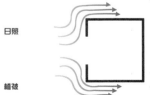
室内要通风，必须有出风口和进风口，这个空间只有一个开口，即使风直接吹也无法通风。

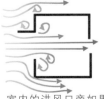
室内的进风口旁如果装了导风板可以增加室内的通风效率。

如果室内有与气流垂直的隔间墙，会妨碍通风。

室内的隔间墙如果与气流平行，则可促进通风。

金陵风吟
Wind Sing of Jinling

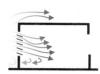 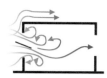  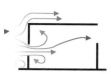

如果进风口有百叶窗，虽然能引导空气流通，但也有可能阻碍通风路径。

如果进风口有旋转窗，将其角度调往室内侧斜向下时，那么空气便会由下而上、涵盖人体的活动范围并从出风口出去。

如果进风口有旋转窗，将其角度调往室内侧斜向时，那么空气便只会在室内的上半部流动，无法涵盖人体的活动范围。

如果进风口低、出风口高，那么就能在人体的活动范围内有效通风。

A shutter at the air intake would guide air circulation and may possibly hamper ventilation path.

If there is a pivoted window at the air intake, the angle thereof is adjusted to be inclined and downward towards the interior space, so the air will cover the range of activity of human bodies from bottom to top and then get out from the air outlet.

If there is a pivoted window at the air intake, the angle thereof is adjusted to be inclined towards the interior space; the air will only flow in a upper portion of the interior space and cannot cover the range of activity of human bodies.

If the air intake is low and the air outlet is high, the effective ventilation will be achieved in the range of activity of human bodies.

一层平面
Ground Floor Plane

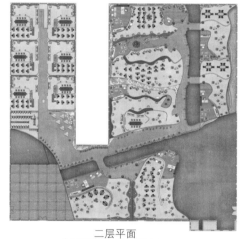

二层平面
Second Floor Plane

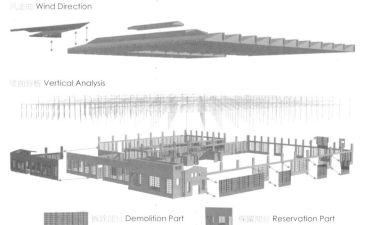

风走向 Wind Direction

竖向分析 Vertical Analysis

拆除部分 Demolition Part　　保留部分 Reservation Part

在人们生态环保意识日益增强的趋势下，自然风的应用已经成为房屋建筑设计当中一项重要的内容，越来越受到人们的重视。

As people's ecological environmental consciousness intensities, the application of natural wind has become an important content of house building design and receives attention of people more and more.

自然风是种廉价、环保的自然资源，建筑师们应该在建筑设计中尽可能利用自然风，降低能耗，减少污染，走符合可持续发展原则的建筑创作之路，不断为保护好自然环境做出应有的贡献。

The natural wind is a cheap and environment-protective natural source. Architects should utilize natural wind as far as possible in the building design and reduce energy-consumption and pollution, and take the path of architectural creation conforming to the principles of sustainable development and continuously make contribution to well protection of our natural environment.

工业遗产保护与室内设计 Industrial Heritage Protection and Interior Design

金陵风吟 Wind Sing of Jinling

电商类办公区
E-Business Office Areas

除了利用柱子模糊空间的功能性，还可以利用大空间中的可变达到为不同职业人群服务的目的。具体操作手法是在办公空间中加入可移动的格挡与导轨系统，可依据使用需求围合出不同属性。

Beside the use of pillars to blur functionality of the space, the variable big space can also be utilized to serve for different occupational populations. The specific operating means is to add the movable blocks and rail-guiding system to the office space and to enclose different attributes according to demands for use.

模式一：部分封闭
Pattern I: Partially enclosed

营造集体空间中的私密空间，可供召开重要会议时使用。

Constructing a private space in the collective space provides use for the important conference.

模式二：全封闭
Pattern II: Totally enclosed

全封闭格挡系统可将预设空间全部围合，营造私密性空间。

Totally enclosed block system can enclose the whole presetting space to construct a private space.

模式三：个人工作室
Pattern III: Personal studio

提供给人数不多的工作室一个互有联系又相互独立的空间。

To provide an interconnected and independent from one another space for the studio having several people.

模式四：团队办公
Pattern IV: Team working

团队办公可以通过不同半径面积的组合达到团队合作的使用目的。

The team working can achieve the use purpose of team cooperation by a combination of areas of different radiuses.

Industrial Heritage Protection and Interior Design　工业遗产保护与室内设计

金陵风吟 Wind Sing of Jinling

风力分析
Wind Analysis

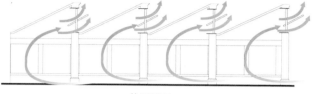

热压通风
Thermal Pressure Ventilation

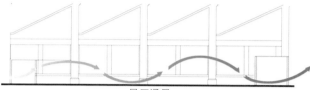

风压通风
Wind Pressure Ventilation

自然风是种廉价、环保的自然资源，建筑师们应该在建筑设计中尽可能利用自然风，降低能耗，减少污染，走符合可持续发展原则的建筑创作之路，为保护好自然环境做出应有的贡献。

The natural wind is a cheap and environment-protective natural source. Architects should utilize natural wind as far as possible in the building design and reduce energy-consumption and pollution, and take the path of architectural creation conforming to the principles of sustainable development and continuously make contribution to well protection of our natural environment.

1	1 A2 一层走廊 A2 Ground Floor Corridor
2　3	2 A2 一层健身房 A2 Ground Floor Gym
4	3 A2 二层小型办公室 A2 Second Floor Small Size Office
	4 A2 二层大型办公室 A2 Second Floor Large-Scale Office

117

金陵风吟
Wind Sing of Jinling

在人们生态环保意识日益加强的趋势下，自然风的应用已经成为房屋建筑设计当中一项重要的内容，越来越受到人们的重视。

As people's ecological environmental consciousness intensities, the application of natural wind has become an important content of house building design and receives attention of people more and more.

1	2
	3
4	5

1 A2 一层个人工作室 A2 Ground Floor Personal Studio
2 A2 二层小型办公室走廊 A2 Second Floor Small Size Office Corridor
3 A2 二层小型办公室走廊 A2 Second Floor Small Size Office Corridor
4 A2 二层小型办公室走廊 A2 Second Floor Small Size Office Corridor
5 A2 二层大型办公室 A2 Ground Floor Ladder Lecture Hall

学生感想
Students' Thoughts

很荣幸参与了此次毕业设计比赛。首先我们想要感谢的是一直孜孜不倦教导我们的老师，其次是感谢这次比赛的组办者，感谢能在大学的尾声画下一个圆满的句号。这次毕业设计让我们受益匪浅：学习了建筑遗产保护的相关知识以及南京城古今文化的绚丽篇章；体会到团队合作的重要性与必要性；明白了应该如何阐述自己的设计想法和如何与伙伴沟通产生的分歧；最重要的是与其他学校同学进行答辩交流时，看到自身的不足与优点，并且吸收来自各大高校的老师、专家的建议和教导，提高我们的设计水平；在毕业设计比赛中不断成长。虽然最终成绩还没有很理想，但是参与过程中我们付出的努力与学习到的东西是最重要的！

It's my honor to take part in this graduate project race. First, we should express our thanks to our hard-working teachers who directed us. Besides, we should thank the race organizers. We are thankful that we have a perfect ending near the end of the university days. This project benefits us a lot. We've learned the knowledge about the architecture relic protection and the Nanjing's glorious culture from ancient times to present. What's more, we experienced the importance and necessity of the teamwork, and understood how to express our own thought and communicate with partners for the objection. Above all, we found our advantages and shortages when debating with student from other colleges, and accepted the suggestions and directions from colleges' teachers and specialists to improve our design levels. We kept growing during the graduate project race. Although we didn't get an ideal result, it's more important for us to learn something through our efforts.

The project use wind as the concept clue. Aimed at energy conservation and modeling, the author tried to rebuild the A1, A2 buildings in Nanjing Chenguang 1865 Innovation Campus. He concluded the site environment, analyzed users' psychological behavior and nature ecology, and then put forward the "the Voice of Breeze in Jin Ling" concept. The concept is uniformly applied to the heritage protection and indoor & outdoor environment design. It shows the integration of wind streamline, element, site function and form. The express of the concept is clear and completed, and the space organization is varied, meanwhile the designer also tried to adapt to the future business. It shows the author's strong professional foundation. Although the separation of A1 exhibition hall and process of guide system are fresh, it may be more reasonable if the author could consider more about the fitness of the exhibition hall and intuition of guide system.

——Huang Yan

教师点评
Teachers' Comments

方案以风为概念线索，针对南京晨光1865创意产业园A1、A2建筑的节能、造型问题进行尝试性改造设计，通过对场地环境的梳理，使用人群的心理行为分析及自然生态等方面研究，提出了"金陵风吟"的设计理念，并在建筑遗产保护与室内外环境设计中统一运用，体现了风的流线、元素与场地功能、形式的有机融合，概念表达比较清晰完整，空间组织富于变化，同时也对未来业态进行了针对性设计，表现出较强的专业功底。而在A1展厅的分隔及导视系统的处理上虽较有新意，但如能从展厅的适宜尺度及导视系统的直观性出发来作更多考虑可能设计合理性更强。

——黄焱

专家点评
Experts' Comments

"金陵风吟"的作者，以"绿色、降耗、节能"为主题，对本题由浅入深的展开设计，他们选择了自然通风做为调节室温的主要手段，首先对南京地区的气候特点及该建筑物的气流都进行了详尽的分析，为下一步具体设计做好技术上的准备。同时通过双层幕墙的设计，恰当控制进、出风口的开关，利用"温室效应"和"烟囱效应"进一步降耗、节能。同时设计者还增加了绿化面积，并在A1、A2之间增加了冥想堂，不仅创造了新的空间，也进一步增强了通风效果。上述所有的设计，都是围绕着"绿色、降耗、节能"的主题进行的，说明了该组同学的设计思路非常明确，并选择了恰当的设计手段来实现这一主题。

——李书才

The author of the "Wind Flowing in Jinling" used "Greening, Lower Consumption, and Energy saving" as the theme. The design work is unfolded from easy to difficult. They choose natural ventilation as the main method of temperature adjustment indoors. First of all, the author analyzed the Nanjing's climate character and airflow around the building in detail in order to prepare for the detailed design in the technique aspect. Meanwhile, the switches of the air outlet and air intake were controlled correctly according to the double-skin facades design. The designer made the most of the "Greenhouse Effect" and the "Chimney Effect" to decrease the consumption and save the energy. He also increased the green area and added a meditation house between the A1 and the A2. It did not only create a new space, but also strengthen the ventilation effect. All the measures of the project were taken around the theme "Greening, Lower Consumption, and Energy Saving". It shows that the thoughts of the group members are very clear, and they chose a proper method to make the theme come true.

——Li Shucai

工业遗产保护与室内设计 Industrial Heritage Protection and Interior Design

CIID "室内设计 6+1" 2015（第三届）校企联合毕业设计
CIID "Interior Design 6+1" 2015(Third Scssion)University and Enterprise Joint in Graduation Design

三等奖
Third Prize

城墙上的老机器
城墙下的老手艺
Old Machine on the Wall, Old Craft under the Wall

高　　校：北京建筑大学
College：Beijing University Of Civil Engineering and Architecture
学　　生：顾文博　付晨辉　王罂伟　余祺盈
Students：Gu Wenbo　Fu Chenhui　Wang Zhaowei　Yu Qiying
指导教师：杨琳　朱宁克
Instructors：Yang Lin　Zhu Ningke

关联页：120-125、213-215

顾文博 Gu Wenbo　付晨辉 Fu Chenhui　王塱伟 Wang Zhaowei　余祺盈 Yu Qiying

平面生成
Plane Creation

在平面图设计上，纳入经济指数考量，采用中国传统城市设计，九经九维，构造平面双核体系，将古城镇植入其中，移步换景，回转道藏。重点刻画公共空间，使A2富有活力。

Add the consideration of economic index into the design of a planar graph, using Chinese traditional urban design, nine by nine dimensional construct planar, to create a bi-nuclear structure system plane, and then embed the ancient town to show a different view with every step, and a different road with every rotation. Focus on public space, to rich the A2 full of energy.

立面示意
Elevation

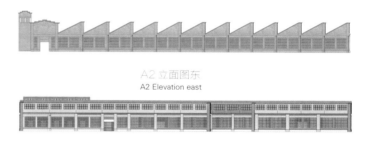

A2 立面图东
A2 Elevation east

A2 立面图南
A2 Elevation south

平面演化
Plan Evolution

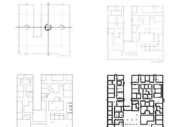

街头巷尾，难见剃头匠身影；
门前屋后，再无挑担子货郎

如今，有多少民进传统艺术正在逐渐流失掉？那些曾经美好的事物，在当下被逐渐遗忘了。可是，还是有人在坚持，是那些承袭了代代相传的独特技艺的手艺人。毕竟这是承载着祖祖辈辈的感情所流传下来的技艺。

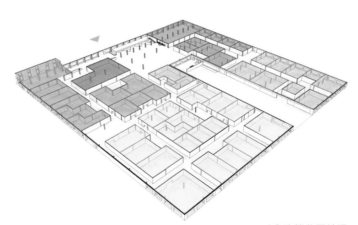

A2 功能分区关系

Nowadays, how much traditional folk craftsmanship is gradually losing? Those things which were beautiful at one time are gradually forgotten by people at moment. However, there are still some people who insist on, who inherited the unique skills of the craftsmen be handed down from generation to generation. After all it is the craftsmanship with generations of feelings.

工业遗产保护与室内设计 Industrial Heritage Protection and Interior Design

分区图
Analysis Diagram

平面分析
Plan Analysis

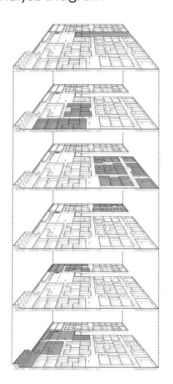

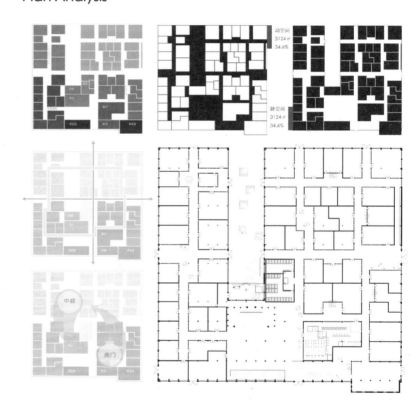

材质解读 Material interpretation

Industrial Heritage Protection and Interior Design　　工业遗产保护与室内设计

A1、A2 人流来源分析
A1、A2 Source of Stream of People Analysis

大报恩寺与园区大门的双股人流形成系统。由此A2中庭为其核心节点。重点刻画，吸引大报恩寺的人流。

A system is formed by double flow of the Dabao'en Temple and park gate. Therefore the A2 atrium is its core node, which is used to attract people flow of Dabao'en Temple.

A2 分区爆炸分析
A2 Partition Exploded Diagram

业态分析
Format Analysis

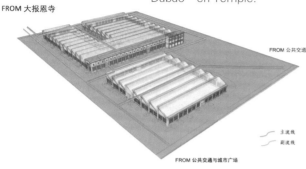

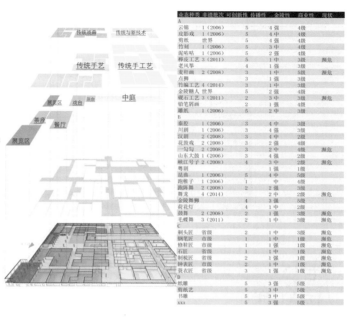

A1、A2 人流来源分析
A1、A2 Source of Stream of People Analysis

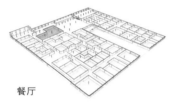

特色民国餐厅，采用棕色的再生木材与民国灰的碰撞，使在二层用餐的人如同置身于南京城墙之上，被限制的视线增加了张力和探索的欲望。而再生木材的选择也中和了原有兵工类建筑的冰冷感。

Characteristic Republic of China restaurant, using the brown recycled-wood match the gray color of old times to make people who have dinner on second floor feel like staying on top of the city wall of Nanjing, make restricted sight to feel more tension and desire of exploration. And the recycled-wood also can balance out the sense of cold of ordnance building.

光膜空间效果
Optical Mode Space Perspective

戏台

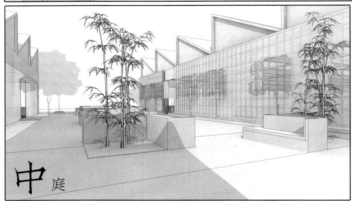

中庭

南门茶座

学生感想
Students' Thoughts

这次毕业设计是一次难得的经历。我们与另外6个兄弟院校的同学们一起切磋专业技能，以技会友。我们学到了许多，体悟了许多。这是我们第一次接触工业遗址的改造设计。一个学期下来，我们认识到了它的难度与重要性。在今后的生活工作中，我们也会牢记此次难得的经历，为了将父辈祖辈珍贵的民族精神、文化遗产传承下去而不断精进。

This graduate project is a rare experience. We made friends with students from other six colleges by the professional skills, and learned from each other. We've learned a lot, and experienced a lot. This is my first time to get in touch with the industry relics reconstruction design. After one semester, we realized its difficulty and importance. In the future, we will keep this rare experience in our mind. We will spare no effort to inherit our ancient national spirit and culture relics.

教师点评
Teachers' Comments

本次毕业设计选题，限定性很强，又具相当的开放性。因此，同学所具有的专业素质和个人的特质都能够充分得表现出来。北建大这组"城墙上的老机器，城墙下的老手艺"作品，较为充分地反映了学生对中国传统空间构成关系的理解，也体现了他们对现代几何美学的钟爱。其设计思路清晰、手法统一。

该作品包含了博物馆、室内创意园区和室外环境改造3个部分，作业量饱满，内容较为丰富。他们遵循了中国传统空间营造法则，以分明的经纬网格来分割空间，以人流动线和空间相互关系所形成的轴线关系来组织空间，并以前厅、中庭、戏院、以及绿化、水体等来形成空间节点、亮点。布局上显得条块分明，区域感、次序感都比较强。室内用材侧重于思考现代人的生活方式和审美趣味以及与周围建筑的融入感，选择了木材、石料、麻布等，力求朴实、自然的效果。

设计在系统性和深入性方面还需进一步加强。如博物馆设计，应更加体现围绕主题思想性和受众互动性的空间视觉与技术表现。

——朱飞

Their favorite in modern geometry aesthetics, the design thought is clear, and the express is uniform.

The work contains museum, inner innovation campus, and outside environment reform. The work amount is full, the content is abundant. They follow Chinese traditional room construction rules; use longitude and latitude net to divide the space, organize space according to the axis relations formed by people streamline and space interrelation, and form the space nodes and highlight according to the antechamber, atrium, theatre, greening, water and so on. The layout is clear, the area sense and order sense are strong. Thinking about the modern life-style, aesthetic interest, and surrounding buildings integration sense, the designers choose woods, stones, and linens and so on to make people feel simple and natural.

The Project should be more systemic and deep-going. For example, the museum should be more ideological to express the theme. Besides, it should be more interactive to the audience, and the technical effect should be more expressive.

——Zhu Fei

专家点评
Experts' Comments

面对特定历史文化场所的设计课题，往往需要"思"多于"做"。即在下笔设计之前，能够花时间在地理、空间、文化溯源、工艺甚至气候条件、人的活动方式等方面做分析、观察与研究。这些设计之前的"准备"工作至关重要。因为缺少了这方面的关注，去做一个方向上欠考虑、追求形式的设计是没有意义的。

这个组的设计应该说在前期思考上是花了功夫的，从南京城市发展、城墙演进与工业遗址的兴衰都做了调研。难能可贵的是，关注到了濒临消亡的传统手工艺。空间的场所与恰如其分的内容结合非常之关键，昔日的兵工厂能够成为培育民间手工艺的场所，又融入当下生活文化，这是一个值得肯定的思路。"城墙"与"手艺"锁定了南京的特征，融入到昔日"兵工厂"遗址中再生。

景观设计中，引入了"护城河"概念的水体，对于活化老建筑区过于萧条、沉闷的气息是有帮助的，能够促进场所转型。如果说更高的要求，则在设计上需要更好地聚集，从众多的分析中理出主线，并且坚决深入去执行。室内与展陈更加有创新与特征地去响应题目"城墙上的老机器，城墙下的老手艺"。

——孙建华

Facing the design project of specific historical site, we need more "think" than "do". Namely, before the actual design, more time should be spent on the analysis, observation and research of geography, space, cultural background, construction craft and even climate as well as people's behavior manners. The "preparation" before the design is necessary. The sole pursuit of design in form is meaningless regardless of the concern on the aspects above.

I can infer that the designers had abundant consideration before the design as they researched on the municipal development of Nanjing, evolution of walls and vicissitude of industrial sites. It is worthy praising that they noticed the endangered traditional handicraft. The integration of site of the space and the appropriate content are vital. The change from a former weapon factory to a place to spread the traditional handicraft with the integration into modern life deserves approval. The "wall" and "handicraft" have locked the feature of Nanjing and revived in the historic site "weapon factory".

Concerning the landscape design, the concept "moat", as introduced water factor, which can inject the vigor into the cold and depressed old building areas and promote the transformation of the site.

As for higher requirement, the design should be more concentrated; main thread should be sorted out from the analyses while it should be firmly implemented. The interior exhibition should respond the topic "Old machines on the wall, Old handicrafts under the wall" more creatively and characteristically.

——Sun Jianhua

工业遗产保护与室内设计 Industrial Heritage Protection and Interior Design

CIID "室内设计 6+1" 2015（第三届）校企联合毕业设计
CIID "Interior Design 6+1" 2015(Third Scssion)University and Enterprise Joint in Graduation Design

三等奖
Third Prize

文化蛹动机——1865影工厂
Culture Worming Machine——1865 Film Factory

高　　校：	北京建筑大学
College:	Beijing University Of Civil Engineering and Architecture
学　　生：	刘凯南　邰嘉琦　姜帅
Students:	Liu Kainan　Xi Jiaqi　Jiang Shuai
指导教师：	杨琳　朱宁克
Instructors:	Yang Lin　Zhu Ningke

关联页：126-131、190-198

Industrial Heritage Protection and Interior Design 工业遗产保护与室内设计

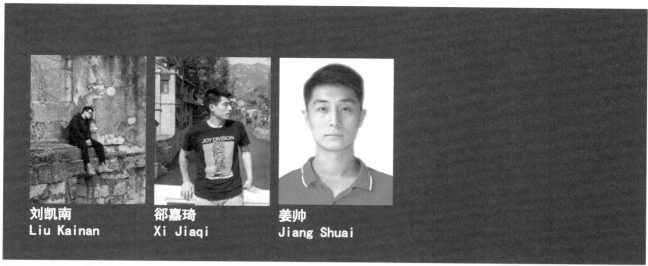

刘凯南 邝嘉琦 姜帅
Liu Kainan Xi Jiaqi Jiang Shuai

设计语汇由来
Origin of Design Vocabularies

不满足于现状的突破。
Not Satisfying with the Status and Getting a Breakthrough.

自从1880年胶卷被发明出来，它就毋庸置疑地成了电影的代名词，但其实中国电影的发展并没有它的形态这么矫情。

Since film was invented in 1880, it has undoubtedly become a pronoun of films. But, actually, the development of Chinese films is not as unreasonable as the morphology thereof.

不断地探寻，翻滚和向上的姿态。
Attitude of Continually Exploring, Turning over and Going Upward.

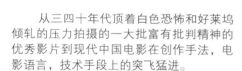

从三四十年代顶着白色恐怖和好莱坞倾轧的压力拍摄的一大批富有批判精神的优秀影片到现代中国电影在创作手法，电影语言，技术手段上的突飞猛进。

From a large number of films with critical spirit produced under the pressure of the white terror and Hollywood dissension at 1930s and 1940s to modern times, Chinese films achieve has been advancing by leaps and bounds on creative practices, film language and technological means.

共享空间 环形空间 下沉空间 封闭空间 开敞空间 虚拟空间 抬升空间

次入口展厅设计
Design of Exhibition Hall of the Secondary Entrance.

此展厅作为摄影机器材的常设展厅，有效过度了兵器展到电影工厂的突兀感，也起到了分流的作用。

This exhibition hall as a permanent exhibition hall of camera equipment effectively transitions an abrupt sense from weapons exhibition to movie industry and also functions to shunt.

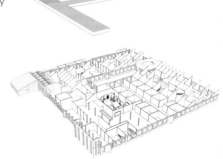

127

产业孵化器
Industry Incubator

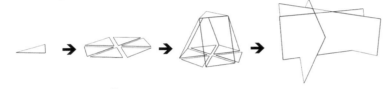

三角形拥有圆形的发散和包容，但同时也有圆所不具有的秩序性。

The triangle possesses characteristics of divergence and containment of the circle, but at the same time, possesses the characteristic of orderliness that the circle does not have.

拥有产业孵化器是一个企业成熟的标志，A2商业定位多为原创品牌或个人作品，而非流水线下的连锁店。为了拒绝绅士化我们引入了孵化器的概念，旨在鼓励、培养、带动新的思想和新的血液。

To have the industry incubator is a mark of a mature enterprise. A2 commerce is positioned on original brands or individual works, rather than chain stores under pipelining. In order to refuse gentrification, we introduced the concept of incubator to aim at encouraging cultivation and bringing along new ideas and fresh blood.

分镜表现图
Storyboard Performance Diagram

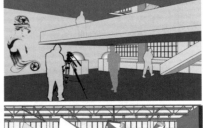

APP 产品及导识概念设计
APP Products and Design of Guiding Concept

导航方面
Navigation

宣传方面
Propagandizing

周边导识与二维码结合（左图）
Combining the Surrounding Guiding with Two-Dimensional Code.

三级导识与二维码结合（右图）
Combining Three-Leveled Guiding with Two-Dimensional Code.

四级导识与二维码结合（下图）
Combining Four-Leveled Guiding with Two-Dimensional Code.

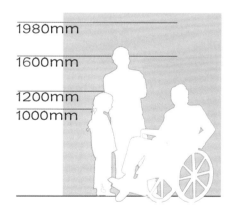
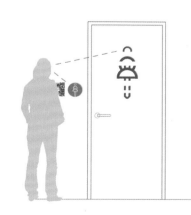

四级导识意向 Intension of Four-Leveled Guiding

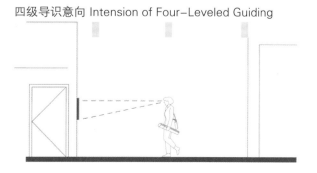

官方公众号"扫一扫"
Official Public Account "Scanning"

网页相关设计
Related Design of Webpage

具有特色的展示照片。
Presenting Photos with Characteristics.

文字介绍 + 链接
Literal Introduction + Link

可切换抬头页
Switch able Page Header

检索标识
Searching Mark

检索标识
Searching Mark

永远跟随 LOGO 及检索条
Following LOGO and Search Bar For Ever.

充足链接
Sufficiently Linking

用户对接
User Butt Joint

方案一采用了 POP 布局模式，整体效果注重用户体验与分享，适合长时间浏览。
Scheme I: Adopting POP layout pattern, the overall effect focuses on users' experience and sharing, which is suitable for long-time browsing.

方案二同样采用了 POP 布局模式，而整体效果更注重于用户的直观使用，简约大气，表面上丧失了体验，但通过点击会别有洞天。
Scheme II: Likewise adopting POP layout pattern, but the overall effect focuses on intuitive use of the user, which has a simple atmosphere and loses experience on the surface, but has a hidden but beautiful spot by clicking.

学生感想
Students' Thoughts

经过了这次长达近半年的六校联合毕业设计，对于即将结束本科阶段学习的我们来说是一次给自己设计能力质的飞跃的最好契机。

在设计伊始，面对着旧工业遗址改造这一既陌生又刺激的题目，我们展开了对北京地区工业遗址的勘探（如798工厂、首钢工业园等地）。

在设计方案阶段，我们构想了一个以影视为主题的方案设计，为了解决概念如何站得住脚以及设计如何做的让人接受这些问题我们力争诠释的完善。

在设计形成阶段，为了更好的表现，我们精心布置了场景以达到效果的完美。

最后，当然，感谢耐心指导的老师，以及并肩作战的同学们。

According to the experience of the six college joint graduate project for half a year, it's the best opportunity for us to improve the design abilities essentially before we finish the undergraduate course.

In the beginning, facing the old industry relic reconstruction project which is unfamiliar and exciting for us, we started to invest the industry relics in Beijing, such as: the 798 Factory, the SHOUGANG Industry Campus and so on.

During the design period, we made a movie theme project. Meanwhile, we spared no effort to make the concept reasonable and tried to make the project acceptable. We tried our best to answer these questions perfectly.

During the formative period, for the better expression, we arranged the scene carefully to make the effect perfect.

At last, thanks for the patient teachers, and brothers fighting together.

教师点评
Teachers' Comments

设计者首先对1865产业园的基地现状进行了深入的分析，并寻找出解决本案的方向。从文化与历史的传承入手对A1展厅进行定位，以城市精神、历史态度、批判反思3点同时切入，结合现代高科技的展示手段设计了一个以电影为主题的轻武器博物馆。

这个通过与武器相关的电影展示来介绍轻武器的设计思路具有独特的创新性。整体设计合理的运用了平时学习的理论知识和设计原理，紧紧抓住历史与电影、历史与文化、历史与建筑艺术这3条主线。

在展厅的功能布局上，分区合理，流线清晰，充分体现了人性化设计。在空间规划上抓住了电影元素的精要部分，通过场景再现的展示手法，使观者意随境转犹如置身于电影中，最大限度的激发了参观者的想象力。

——刘晓军

Firstly, the designer analyzed the current situation of the 1865 Industrial Park Base carefully, and tried to find the way to finish the case. He started with the inheritance of the culture and history to position the A1 Exhibition Hall. Based on the city's spirit, history attitude and critical reflection, and combined with high-tech presentation, he designed a light weapon museum whose theme was the movie.

It is particularly innovative to introduce the light weapons by the related movies. The designer reasonably made the most of the knowledge of the theory and principle, and firmly grasped three lines: history & movie, history & culture, and history & architecture.

In the exhibition hall function layout aspect, the layout is reasonable, and the streamline is clear. It is fully reflect the humanize design. During the space planning, it mastered the essentials of the movie factor. By recovering the scene, it could make the audience feel that they are in the movie, and furthest motivated their imagination.

——Liu Xiaojun

专家点评
Experts' Comments

文化蛹动机——1865影工厂。在一个建筑项目与城市精神之间找到确切的关联，是该项目设计者的出发点。"生生不息，坚韧不拔"，不仅是一座城市的精神，更可以具体体现在一个工厂的历史变迁之中。设计团队对项目的了解比较深入。展馆内斜向的布局与动线，加强了动态感受。拟定的展览主题虽然不够清晰严谨，但对于观众参与互动的思路是正确的，每一段历史的呈现以电影影像的形式表达出来，生动而有活力。

——陈卫新

The cultural perpetual motion machine——1865 Movie Factory. The purpose of the project designer is to find out the exact relations between a construction and the city's spirit. "life and growth in nature, grittiness stick-at-itiveness". It's not only a city's spirit, and it is reflected in a factory's historical changes in detail. The design team members understand the project deeply. The slanting layout and the kinetonema in the exhibition hall strengthen the motivate feeling. Although the proposed exhibition theme is not clear and precise, it's correct to let the audience take part in the interaction. Each of the historical stories is expressed by the movie image. It's vivid and dynamic.

——Chen Weixin

工业遗产保护与室内设计 Industrial Heritage Protection and Interior Design

CIID "室内设计 6+1" 2015(第三届)校企联合毕业设计
CIID "Interior Design 6+1" 2015(Third Scssion)University and Enterprise Joint in Graduation Design

三等奖
Third Prize

时 · 境
Feeling · Space

高　校：	浙江工业大学
College:	Zhejiang University of Technology
学　生：	夏欣
Students:	Xia Xin
指导教师：	吕勤智　黄焱
Instructors:	Lv Qinzhi　Huang Yan

关联页：132-135

Industrial Heritage Protection and Interior Design 工业遗产保护与室内设计

夏欣
Xia Xin

设计说明
Design Notes

我的设计以时间、空间为线索，进行时间的推移和空间的延展，形成老时光，新境物的构想，最终得到了"时境"这个概念。时境，既是人们对南京城市中老建筑有所触动的心理感受，也是老建筑在一定时间维度里的自我塑造的能力。在顺应时代的大背景下，剪去人们的记忆断层，重新连接起人们对老建筑那温和而又充满记忆的空间情绪。

The concept of "Feeling space" was finally obtained by my design which it was based on time and space for clues, with the passage of time and extension of space, to form an old time and the new environment. Feeling Space is not only the psychological feelings that people have for the old building of Nanjing city, but also the self-fashioning ability that old building has at a certain time dimension. On the premise of conforming to the times background, cutting off our part of absence of memories and reconnect our sense of mild and memories about old buildings.

设计定位
Design Position

艺术家聚集的空间往往是能营造开拓自己的价值空间，也是能够提供艺术创作来源的空间，也是能够张扬个性自由、有弹性的空间。

The Space which artists could get together is also the space which can develop their value, which provides source of artistic creation, and also the space that is able to demonstrate individually with freedom character and flexibility.

场地问题
Field Problem

使用者 Users　　感官 Senses　　空间与需求 Space and Requirements

（时间）流逝 单调 平静 寻找新起伏
(Time) Passing Monotonous Peace Looking for new ups and downs.

Artist Studio

（空间）固定 单一 乏味 寻找新风景
(Space) Fixed Single Boring Look for new view.

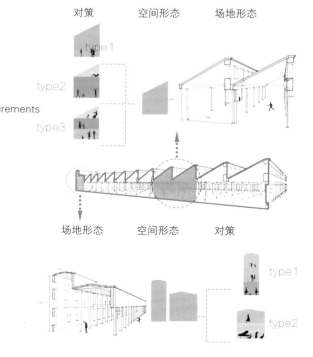

工业遗产保护与室内设计 Industrial Heritage Protection and Interior Design

设计概念
Design Concept

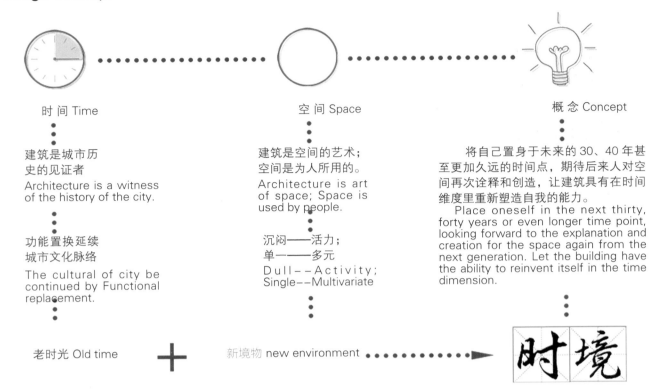

时 间 Time

建筑是城市历史的见证者
Architecture is a witness of the history of the city.

功能置换延续城市文化脉络
The cultural of city be continued by Functional replacement.

老时光 Old time

空 间 Space

建筑是空间的艺术；空间是为人所用的。
Architecture is art of space; Space is used by people.

沉闷——活力；
单———多元
Dull--Activity;
Single--Multivariate

新境物 new environment

概 念 Concept

将自己置身于未来的30、40年甚至更加久远的时间点，期待后来人对空间再次诠释和创造，让建筑具有在时间维度里重新塑造自我的能力。
Place oneself in the next thirty, forty years or even longer time point, looking forward to the explanation and creation for the space again from the next generation. Let the building have the ability to reinvent itself in the time dimension.

时境

设计分析
Design Analysis

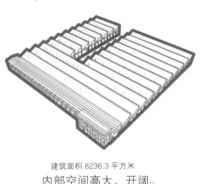

建筑面积 8236.3 平方米
内部空间高大、开阔。
Inner space is tall and open

朝北向的锯齿形屋顶
锯齿形屋顶形成的三角形空间利用率不高。
The ratio of triangular space formed by a zigzag roof is not high.

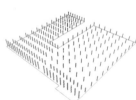

规则布置的柱网
柱网间距宽敞排布规则且密。
Column with wide spacing but concentrated arrangement.

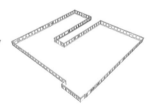

保存完好的历史外墙
民国建筑保存完好为历史保护性建筑
The architecture of the republic of China is well preserved as historic conservation building.

分割空间 组织交通
划分大小空间规划流畅交通
Dividing the size of space and planning smooth traffic.

分割二、三层空间
充分利用高大开敞空间
Make full use of large and open space.

置入单元空间
保证空间使用的多样性
Space diversity.

分隔二层空间
多样的展陈方式
A variety of display modes.

学生感想
Students' Thoughts

毕业设计阶段的学习到此已画上了句号,感谢此次校企联合毕业设计为我们提供了一个多校交流互相学习的平台,感谢各位老师的指导与鼓励,让我在短短数月的学习中收获良多、感触颇多。

到毕设结束,整个毕业设计过程下来,从最初的课题认知、个人的思考到初期方案的呈现到最终方案的敲定与深化,比起最后的结果,我更享受的是自己挑战整个毕设的过程,虽然过程中时而反复否定的自我及方案,时而方案卡壳纠结进行不下去,但我仍很享受这个过程带给我的成长与酸甜苦咸的感受。

继续吧!在路上,寻找正能量,去做最好的自己。

The study of the graduation design has ended. Thank you for the joint of college enterprises graduation designs providing us with a communication platform of multi-colleges to learn from each other. Thank you for the guidance and encouragement from the teachers let me gain and feel a lot during the months of study. In the process of the design, from the original project cognize and individual thoughts to the presentation of initial plan, and then to the confirmation and deepening of the final plan, I prefer challenging it by myself to reach the final result. Although I denied the plan or even myself repeatedly during the design, and sometimes it was stuck without going on, I still enjoyed the process bringing me the growth and feelings of sour, sweet, bitter and hot.

Go on! To look for positive energy on the road, and to do the best you can.

教师点评
Teachers' Comments

建筑是动态的事物而不是静态的,建筑随时在变,该方案在顺应时代的大背景下,重赋老建筑生命。从满足不同使用者的个性需求、空间应对时间的变化、增加使用者的参与度三个维度展开设计。

方案设计之初,深入剖析了A2建筑的围护结构、建筑特征、详细尺寸等因素,归纳总结了目前存在的主要问题。明确该空间的角色定位,提供一个艺术家聚集的场所,由此展开空间设计,提供艺术创作来源的空间,也是能够张扬个性自由、有弹性的空间。

空间设计手法是该方案的重点,针对不同的空间尺度,划分不同的空间形式,丰富使用者的不同空间尺度体验。详细探索了4种工作空间的可能性及具体空间尺度。展示空间强调空间的灵活性、开放性、连续性,设计具有很强的创新性思考。

——朱宁克

The building is the dynamic thing instead of the static one. It remains changing at any time. The project conforms to the historical background, and the old building is endowed with a new life. The design work was started with three dimensions: various individual needs, transform of the space adapted to the time, increased user's participation.

In the beginning, the designer analyzed the factors of the A2 building carefully, such as: the building enclosures, building features, detailed measures of the building and so on. Then he concluded the main problems at present. It gave the space a clear role: to provide a place for artists to gather together. Then the designer started to unfold the design. It is the elastic space for the artist to provide the source of the art creation, and express individual freedom.

The main point of the project is the design method of the space. According to different space dimensions, it was divided into different space forms. It could enrich the users' experiences of different space dimensions. The designer carefully explored the possibilities of 4 working spaces and their space dimensions in detail, showed the space and emphasized its flexibility, openness, and continuity. The project is filled with strong innovative thought.

——Zhu Ningke

专家点评
Experts' Comments

作品"时境",创作者对建筑的活化利用着眼点在"空间",强调空间与时间与使用者的关系,空间因有了人(主要指艺术家)的参与以及时间维度的变化而鲜活了、饱满了,原有空间变戏法般的延展出尺度多样界面丰富的各类功能空间,旧址不再是大的厂房尺度和较为单一的空间,街区化、生活化、艺术化了,这为课题的研究带来新的思路。

创作者层层分析与推导,但"分析问题"并未能较好的作用于"解决问题",对旧建筑美学特征和空间特质的如何妥善利用缺乏交代,而所呈现的成果由于过于依赖"艺术家"群体和未来的参与,反而难以阅读到"此时此地"空间的魅力与价值。

——邹超

The work "Time & Environment", the author focused on "space" to make the most of the activation of the building. He stressed the space and its relation with time, and relation with users. It makes the space become active and full because of the people's participation and changes of the time dimension. The previous spaces were changed magically into all kinds of function spaces with various measures and abundant surfaces. The site is no longer the single space with factory's measure. It becomes a lifestyle and artistic block. This is a new thought for the project.

The designer analyzed and deduced the problems layer upon layer, but "analyzing the problem" didn't act on "solving the problem". He didn't clearly explain how to take advantage of the aesthetic features of the old buildings and their space qualities. Because the efforts rely on "artist" group and future participation seriously, it's hard to find out the charm and value of the space "right here right now".

——Zou Chao

工业遗产保护与室内设计 Industrial Heritage Protection and Interior Design

CIID "室内设计 6+1" 2015（第三届）校企联合毕业设计
CIID "Interior Design 6+1" 2015(Third Scssion)University and Enterprise Joint in Graduation Design

佳作奖
Excellence Prize

锯象金陵——从制造到智造
Zigzag Jinling
——From Manufacturing to Intelligent Production

高　　校：	同济大学
College :	Tongji University
学　　生：	周怡　陈杰
Students:	Zhou Yi　Chen Jie
指导教师：	左琰
Instructors:	Zuo Yan

关联页：136-143

周怡 Zhou Yi　　陈杰 Chen Jie

设计概念生成与依据
Design Concept Generation and Evidence

在着手此次设计的时候，我们首先对南京文脉和基地特征的进行分析，将其归纳为锯象金陵。锯象，字面上的意思是锯齿形的，而内在却是曲折而跌宕起伏的，暗合了南京历史文脉发展、地理环境和基地建筑的特征。同时基于对现场的理解和感受，由此提出设计概念，从金陵制造到金陵智造。

首先，我们展开分析锯象金陵，通过百年南京记录片和档案史料的阅读，我们发现南京的历史十分跌宕起伏，它是六朝古都，同样也是民国政府的首要之地，在传统文化与现代文明之间的挣扎起伏，而且南京历史上的重大事件，导致多次人口剧变，这种历史发展的跌宕起伏，是锯象的。

其次，当我们实地走访的时候，发现南京城区起伏不平，市内聚集了许多山，形成山多、水多、丘陵多的地貌特征，这种地势上的起伏不平，是锯象的。接着我们分析出连续性的锯齿形屋面，是基地建筑最首要的特征，能够使其最节约材料，并且获得稳定的采光空间。作为包豪斯建筑，建筑外形必须从社会和技术的前提条件出发，其内核是时代精神。

When setting about the design, we first analyze Nanjing context and base features and then conclude it to Zigzag Jinling. Zigzag literally means saw tooth and has an internal meaning of ups and downs, which impliedly means the feature of historical context development, geographical environment and base buildings of Nanjing. Based on understanding and feeling to the spot, we put forward the design concept: from Jinling manufacturing to Jinling intelligent production.

First, we spread out analysis of Zigzag Jinling: by reading Nanjing documentary in one hundred and file historical data, we find that the history of Nanjing has a lot more ups and downs. Nanjing is an ancient capital of Six Dynasties and likewise a chief place of Government of the Republic China with ups and downs in struggling between the traditional culture and the modern civilization. The important event in Nanjing history results in vast changes of population for many times. Such ups and downs of history development are zigzag.

Second, when we make on-site visit, we find that the urban area of Nanjing is undulating. There are a lot of mountains gathered in the city to form a geomorphic feature of many mountains, more water and more hills. Such undulating terrain is zigzag.Then, we analyze the succession zigzag roof covering, which is the major feature of base buildings. We can minimize the cost of materials of it and obtain a stable daylighting space. As Bauhaus buildings, the architectural appearance must start from the precondition of society and technology and have a core of spirit of the time.

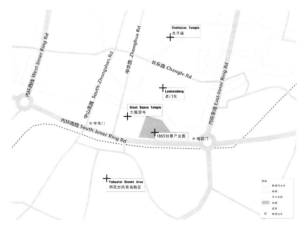

南京1865园区周边环境与基地概况

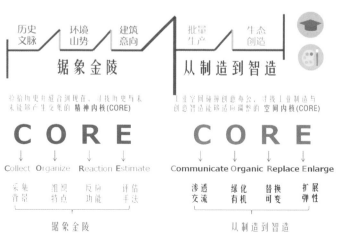

创业孵化器的精神内核与空间内核

工业遗产保护与室内设计 Industrial Heritage Protection and Interior Design

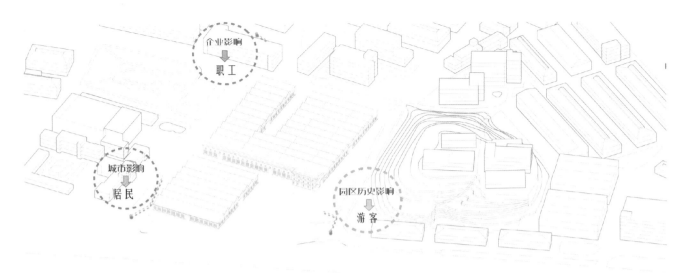

功能引入与平面生成
Function Introduction and Plain Generation

就业问题是年轻人和社会最为关注的，在历经几个最难就业年后，2015年被戏称为"更难就业年"。南京作为每万人拥有在校大学生数量全国第一的高等教育中心，同样面临这样严峻的问题。同时南京自古教育繁盛，当下的321人才培养计划也受到格外重视，由此南京非常看重人才培养。面对如此突出的就业矛盾，文化创意产业相对投入少，门槛低，核心是创造力，非常适合对新事物敏感、有梦想、有冲劲的大学生群体。所以工厂转型成为创意工厂，不仅是产业转型的出路，也为大学生创业者提供了一个施展抱负的平台。

至此我们对这个课题进行思考，在旧建筑改造设计中，是一场捡拾历史缝合到现在的过程，寻找历史与未来能够产生交集的精神内核。作为内核的CORE，也展示了我们做的前期工作内容，即采集背景、组织特点、反应功能和评估手法。以下我们将展开阐述从金陵制造到智造的设计概念，分为三部分：智入功能、智变空间和智慧改造。而在空间设计中，我们将原有理念分四方面探讨工业制造与创意智造能够相互适应调整的空间内核，分别是渗透交流、绿化有机、替换可变和扩展弹性。

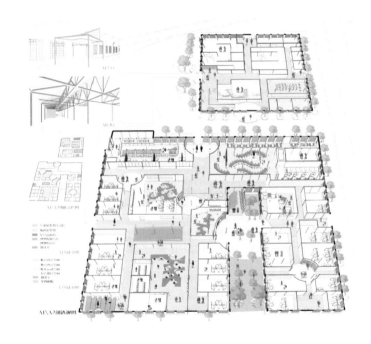

As students to graduate on hand, the problem of employment draws the most attention of young people and society. However, after several hardest employment years, 2015 gains the nickname of the "harder employment year". Nanjing as a center of higher education that is No. 1 to have the number of college students per ten thousand people in the country, also encounters such a severe problem. At the same time, Nanjing has prosperous education since ancient times. The current 321 talent training plan attaches great importance, as such; Nanjing pays abundant attention to talent training. Confronting such outstanding employment conflict, the cultural creative industry has a relatively small input and a low threshold, the core is creativity which is very suitable for college students who are sensitive to new things and have dreams and motivation. Therefore, the transformation from factories to creative factories not only is an outlet of industrial transformation but also provides a platform for entrepreneurs of college students to realize their aspirations.

So far, we think about this project, the reconstruction design of old buildings is a process of picking up the history and then sewing it with now to seek for a spirit core where the history and future can generate intersection. The CORE as the inner core also shows our preliminary work content, i.e. collecting background, features of organization, response function and assessing skill. Hereinafter, we will explain the design concept from Jinling manufacturing to intelligent production, which includes three parts: intelligently embedding function, intelligently changing space and intelligently reconstruction.In the space design, we divide the original concept into four aspects to discuss the space inner core in which the industrial manufacturing and intelligently creative production can adapt to each other for regulation, including penetration communication, greening organic, substitution variable and expanding elasticity.

创意办公模式的研究
Study of Creative Office Mode

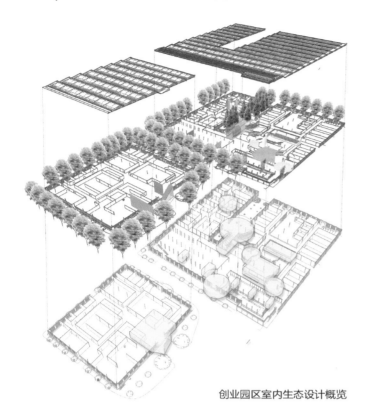

创业园区室内生态设计概览

创业园区室内休憩点

沉淀池改造运动场效果图

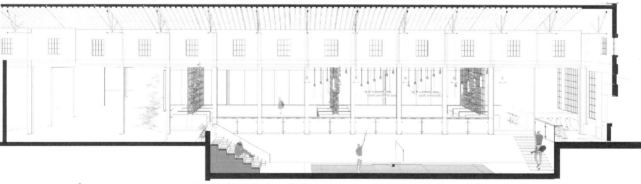

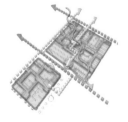

相对于传统的办公模式，创意办公、Loft模式和联合办公（Co-working）模式更加适用于创业起步状态的个人和小团队，提供性价比极高的办公场所与交流性更强的商业环境，成为我们创业孵化器办公模式的首选。

而考虑到建筑的新颖、创意、活力和先锋性，我们将生态理念引入到建筑的公共空间设计中。在汇聚人流的圆厅、内广场、餐厅、娱乐室当中采用了室内绿墙，既能够保证室内的空气质量、湿度与温度的适宜，也为创业人群提供了很好的视觉享受。

而在一些人流交织的公共空间，我们也采用了开放式的Lounge、Bar和Pantry的形式，为创业人群提供优良的后勤服务，同时也提供了社交、商谈的区域，任其产生创业的化学反应。

Relative to the traditional office mode, the creative office mode, loft mode and coworking mode are more adapted to an individual or a small group in a state of start-ups, and can provide the office space with very high cost performance and the business environment with stronger community. It will become a first choice of our business incubator office mode.

In view of novelty, creativity, vitality and pioneer spirit of the building, we introduce ecological idea into public space design of the building. The use of interior green wall in round hall, inside square and restaurant, and entertainment room where the stream of people gathered can ensure suitable interior air quality, humidity and temperature and provide good visual enjoyment for business people.

In some public spaces with crowded people, we also adopt the form of opening Lounge, Bar and Pantry to provide excellent logistics service for business people and at the same time, provide social contact and negotiation areas to make it freely generate chemical reaction of starting a business.

室内生态设计
Interior Ecological Design

我们在设计中引入了大量的生态设计，除了前文提到的室内生态绿墙之外，我们还在一些公共空间乃至办公空间采用了爬绿的形式来增添绿化感受。在A2的室外庭院中，我们除了对原有的绿化采用了保护、保留、优化的手法，还增加了一些绿色步道等，让健康与环保成为创意园区的一大特点。同时，我们在许多室内空间的铺地、墙面装饰上采用了南京老城区（老门东、门西）拆迁时回收的木材、旧瓦等材料，再循环使用使得园区的绿色理念更加深入清晰。

同样地，鉴于南京雨水多的特点和锯齿状屋面集水性极佳的优势，我们也将雨水收集系统引入到建筑之中，通过雨水再循环为建筑中的生态绿化浇灌，更为节能环保。

We introduce a lot of ecological design to our design, besides the interior ecological green wall mentioned above; we also increase the greening feel by adopting the form of green-climbing in the public space and even office space. In the outdoor courtyard of A2, we protect reserve and optimize the original greening and further add some green footpaths, so that the health and environment protection become big characteristics of the creative park. Meanwhile, we adopt recovery materials such as wood and used clay tiles generated when Nanjing old city areas (old door east, door west) were removed on pavage and wall decoration of the interior space to re-use them and make the green concept of the park clearer.

Likewise, in view of the characteristic of heavy rainwater of Nanjing and excellent water-collection of the zigzag roof surface, we introduce the rainwater collecting system into the building to water the ecological greening in the building through recycling of the rainwater, which is much more energy-conservative and environment-protective.

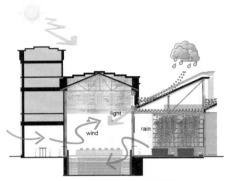
建筑生态循环系统

A2公共空间效果图

A2联合办公空间效果图

在空间的丰富性上，我们在多数区域尊重了建筑的历史风貌，譬如原A2东南角的沉淀池，我们将其高差还原成为一处下沉式的运动场所，并在其中加入Loft元素，使其成为整个设计中最具活力的场所。同时，在少数区域中，我们在不破坏建筑结构的前提下做了空间上的激活设计，譬如圆厅、绿道的设计，改善了工业建筑大体量所造成的沉闷与流线固化问题。

On richness of the space, we respect the historical style and features of the building in majority regions, for example: a sedimentation tank of the original A2 south-east corner, the altitude difference thereof is restored to a sunken sporting site and loft elements are added thereto to make it become the most vital place in the overall design. At the same time, we make activated design on the space without destroying the building structure in a minority area, for example: the design of round hall and greenway, to improve heaviness and flow line curing problem caused by big mass of industrial building.

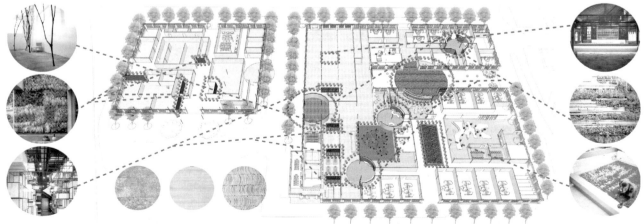
园区设计效果概览

创意空间与联合办公空间设计
Creative Space and Co-Working Space Design

我们在设计中以创意作为主旨，为空间提供更多的便利性与可能性。不同于传统办公空间的固定性、高效性与封闭性，创意办公和联合办公 Co-working 讲求的是高开放度、高参与度、流动性与无限可能性。我们会为创业者提供的不是死板的办公桌椅套件，而是趣味性与实用性并重的工作区域。鉴于创业伊始的创业者没有固定的客户、固定的合作伙伴甚至固定的办公时间，我们将办公空间设计成可随时拎包就座的开放式工作站，类似于图书馆的自习模式。而针对可能有固定办公需求的创业团队，我们将提供流动性稍弱的办公集装箱模式，提高私密度和固定性。

同样地，园区内的餐厅是供创业者就餐的，而在厨闲时间（非营业时间）利用率并不高，所以我们采用了可以灵活组合的餐台，为其提供多种组合模式，可以满足用餐、会议、讲座、开放办公等多种使用需求，提高其利用率。

而对于 A2 创业孵化器的空间排布，我们首要考虑的是 A2 侧企业园区人流的引入，增加创业工作室与既有的大型企业之间的相互可达，从而促进企业平台和创业平台之间的交流。其次我们考虑 A2 内部创业平台空间之间的流线，以及内部人流的空间可达性，为创业者的合作、抱团提供条件。

考虑到不同时期对空间可变的需求，我们设置了一定配比的弹性办公和弹性餐饮空间。可以根据前一年的报表灵活改变不同面积工作室之间的比例，也可以根据活动和会议要求改变餐饮空间的功能。

We take creativity as a gist in the design to provide more convenience and possibility for the space. Different from fixity, high efficiency and closure of the traditional office space, the creative office and coworking lie in high opening degree, high degree of participation, mobility and unlimited possibility. We provide for the entrepreneurs the work regionof interest and practicability instead of stiff office tables and chairs suite. The entrepreneurs of start-ups do not have the constant customer and partners and even the fixed office hours, we design the office space as an open style working station that is available at any time similar to self-study mode of the library. For the entrepreneurial team that may have the demand for the fixed office, we will provide the office container mode with a slightly weak mobility to improve privacy and fixity.

Likewise, the restaurant in the park is to provide meals for the entrepreneur and has a low use ratio at kitchen spare time (non-business hours), so we adopt the dining table that can be combined flexibly to provide several combining modes to satisfy multiple use demands of dining, conference, and lecture and open office so as to improve the use ratio.

As to the space distribution of A2 startup incubator, we first consider the introduction of stream of people in the enterprise park of A2 side to increase the inter-reaching between the startup office and the existing large-scale enterprise so as to promote communication between the enterprise platform and the startup platform. Second, we consider the flow line among the interior startup platform space of A2 and the space accessibility of the interior stream of people to provide conditions for cooperation and huddling of the startup.

In view of variable demands for space at different times, we arrange a certain ratio of elastic work and elastic dining space. The proportion among studios of different areas can be changed flexibly according to report forms of the previous year, and the function of dining space can be changed according to requirements of activities and conferences.

联合办公空间剖面

联合办公空间剖面

联合办公空间效果图（一）

联合办公空间效果图（二）

联合办公空间效果图（三）

创意空间与联合办公空间设计
Creative Space and Co-working Space Design

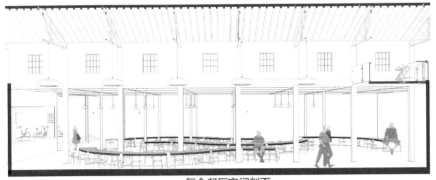
复合餐厅空间剖面

餐厅吧台效果图

餐厅讲座模式效果图

餐厅自习模式效果图

结语
Conclusion

我们希望在南京1865这一极具历史底蕴的园区环境中勃发出一股新的属于这个时代的生命力。在对旧工业建筑保持充分的尊重的前提下，对其进行有节制的、温和的改造利用，延续其辉煌的民族历史，让它的标杆性、前瞻性在当代获得新的定义。

We hope to generate a stream of new life force belonging to this era in the park environment of Nanjing 1865 which has historical sedimentation. With the premise of maintaining sufficient respect to the old industrial building, we make the abstinent and gentle reconstruction to last the glorious national history and thereby make the benchmarking and perceptiveness there of have new definitions at the present times.

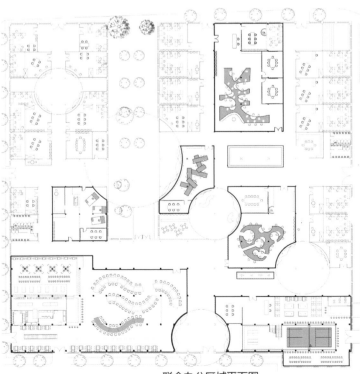
联合办公区域平面图

学生感想
Students' Thoughts

此次毕设与各大院校的同学友人们一同对旧工业建筑的保护与再利用进行了持续而深入的探讨。从前期的开题相聚于南京六朝古都，在烟雨朦胧中体悟金陵的百年神韵；到中期汇报时在西安古城，亲身感受了中原粗犷中的细腻；最后在广州这一代表传统与时尚相互交融的一线城市，通过最直接的视觉与触感比较其与上海都市风格的不同。这些游历的过程为我们的设计提供了大量的灵感，也将成为影响我们一生的宝贵经历。

This project was the cooperation with classmates and friends from all over the colleges. We had a continuous and deep discussion about the old factory building's protection and its re-use. In the early stage, we gathered together in Nanjing, the capital of six ancient dynasties, for the opening speech. We experienced the one-hundred year long spirit of the city in the drizzle. In the mid-term, we reported the progress in Xi'an, a city with long history. We feel the meticulous factors among the rough atmosphere in Central Plain. Finally, we were in Guangzhou, a big city combined with tradition and fashion. We compared the city with Shanghai by vision and feeling. These experiences provide us a plenty of inspiration, and they will be a precious experience that influence us in our lives.

This scheme, starting from the feature of sawtooth-shaped roof of A1, A2 workshops in original Jin Ling manufacturing Bureau, has designed the two historical protection buildings combined with their location advantages as the creative office buildings serving the business incubators for youth in compliance with Nanjing 321 talent cultivation plan after researching on the status of other creative parks in Nanjing and with reference to Nanjing municipal master plan, whose environment has leapfrog development from original weapon manufacture to future intellectual creation. The space design, with full consideration of the position of door and window of the original buildings, column grid and land subsidence, has connected the interior flow lines with several round public traffic nodes and introduced the green wall, rainwater recycling and other energy-saving technologies, which injects the vigor and vitality into the cold and desolate industrial environment. The design theme is definite and the renovation methods are appropriate, representing a wise utilization of the original environment.

——Zuo Yan

教师点评
Teachers' Comments

本方案以南京金陵制造局 A1、A2 两栋厂房建筑的锯齿状屋顶特征为抓手，在对厂区历史变迁、南京其他创意园业态功能调研后，参考南京城市总体规划，将这两个历史保护建筑结合地理区位优势改造成符合南京 321 人才培养计划、面向年轻人的创业孵化器为主体的创意办公环境，实现环境从原先的兵器制造向未来的思想智造的跨越式转变。空间设计充分考虑原建筑的门窗位置、柱网关系及地面沉降等因素，以几个圆形公共交通节点将流线串接起来，引入了绿化墙、雨水循环利用等生态节能技术，给原本萧条冷漠的工业环境重新注入了新的活力和生命。设计主题明确，方法得当，对原环境的利用较为巧妙。

——左琰

工业遗产保护与室内设计 Industrial Heritage Protection and Interior Design

CIID "室内设计 6+1" 2015（第三届）校企联合毕业设计
CIID "Interior Design 6+1" 2015(Third Scssion)University and Enterprise Joint in Graduation Design

佳作奖
Excellence Prize

离 火
Li•Fire

高　　校：	浙江工业大学
College：	Zhejiang University of Technology
学　　生：	马凯杰
Students：	Ma Kaijie
指导教师：	吕勤智　黄焱
Instructors：	Lv Qinzhi　Huang Yan

关联页：144-151

马凯杰
Ma Kaijie

设计说明
Design Description

选了火焰的概念，以一种自由的建筑空间姿态，是一个精神中心，然后去想象到底什么功能有这样作用，围绕着这个精神中心展开的展览，也就是一种新的空间形式，打破了一个空间串联一个空间的方式。

The concept of Fire I chose by a kind of untrammeled pose of architectural space is a spirit center. Then to image that what function has this effect, I regard the exhibition round the spirit center as a new spatial form, breaking the mode of spaces connection in series.

设计理念
Design Concept

火与形式
Fire & Form

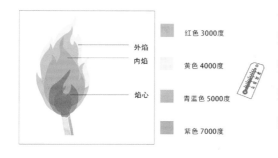

火焰从红色橙色（3000度）　　The fire from red and orange(3000℃)
到黄色白色（4000度）　　　　to yellow and white(4000℃)
到青色蓝色（5000—6000度）　to cyan and blue(5000℃—6000℃)
到紫色（7000以上）　　　　　to purple(above 7000℃)
到最后看不见的紫外线（几万度）to invisible ultraviolet light(Tens of thousands of degrees Celsius)

"以柱为木，空间为火"的概念
The Concept of "Pillar as Wood, Space as Fire"

八卦中的离火
The Li Fire in the Eight Trigrams (BaGua)

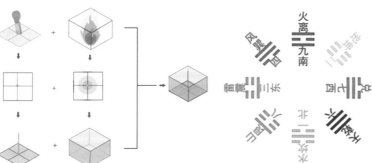

一般说到离火，就是指的五行中的火。八卦之中，离为南方，因南方为太阳正午之位，属阳，且四象之中南方属于丙丁之火，为朱雀守护。所以八卦中离位为火，即离火。

When talking about Li Fire, it means the Fire in the Five Elements (WuXing). In the Eight Trigrams, Li is the south. As the south is the place when the sun is at noon, it belongs to "Yang". In the four images (SiXiang), the south belongs to the Fire of the 3rd and 4th of the ten Heavenly Stems, defended by Suzaku. So the Li palace in the Eight Trigrams belongs to the Fire, called Li Fire.

工业遗产保护与室内设计 Industrial Heritage Protection and Interior Design

理念与建筑结合
Concept & Architecture

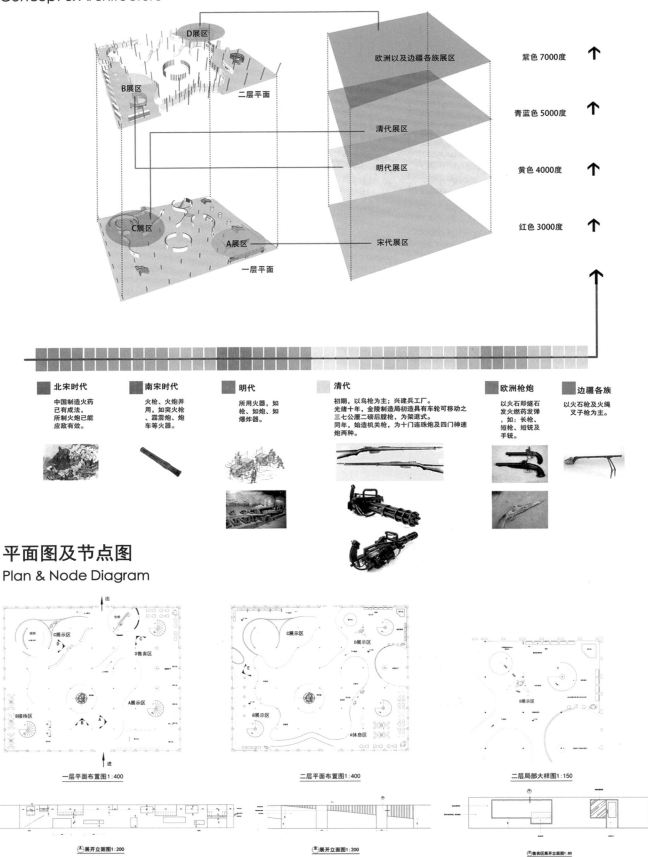

平面图及节点图
Plan & Node Diagram

流线分析
Streamline Analysis

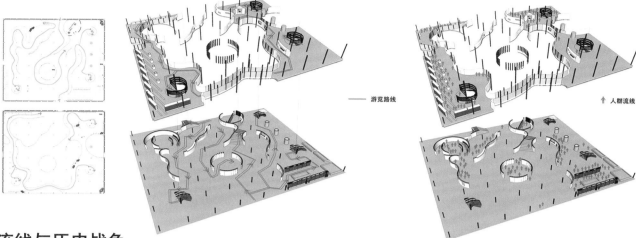

流线与历史战争
Streamline & Wars

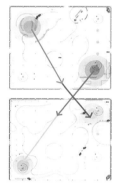

　　火器必定伴随着战争，火器的发展史，必定是战争的发展史。

　　通过对宋、明、清的战事，来诠释火器，看完博物馆不仅仅只是对中国火器史有了了解，更多的是对中国战争朝代历史的一个略览。所以，我对博物馆的定位是战争与火器并存，相辅相成。流线的行进是将战事按照时间顺序，使参观者一步步去了解中国战争史。

Firearms must come from the wars. The history of the firearms must be the development history of the wars.

Interpreting the firearms by the story of Song, Ming and Qing dynasty, you can know not only the history of firearms in China, but also the history of wars of Chinese dynasties. Therefore, I position the museum as a place where wars and firearms coexist and supplement each other. The travel of the streamline put the story in chronological order helping to understand the war history in China.

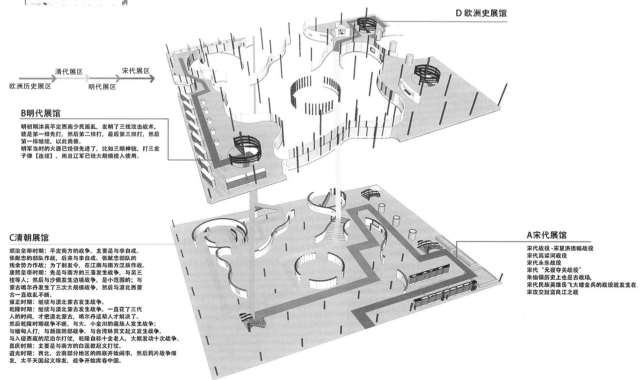

功能分区
Function Division

博物馆一层与二层之间有五个楼梯，三个旋转楼梯与两个大堂楼梯交换上下两个空间。

展厅——休憩，休憩——展厅，展厅——纪念品区，功能与空间通过楼梯连接。

There are five flights of stairs between the ground and the first floor of the museum, including three flights of spiral stairs and two flights of lobby stairs, exchanging the spaces between the two floors.

Exhibition hall to lounge hall, Lounge hall to exhibition hall, Exhibition hall to souvenir area, Functions and spaces connected by the stairs.

空间分析
Spatial Analysis

空间以公共开敞空间、半公共开敞空间、半私密休憩空间三种空间为主，交叉空间存在形式，垂直空间存在形式，让在不同地点的人，能有柳暗花明之感，亦有不疲乏之感。

There are three main kinds of space: public open space, half-public open space and half-private rest space. The cross-space form and vertical-space form offer the sight of dense willow trees and bright flowers and the untired feeling to visitors in different locations.

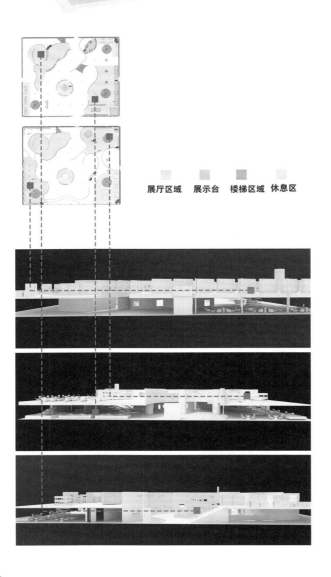

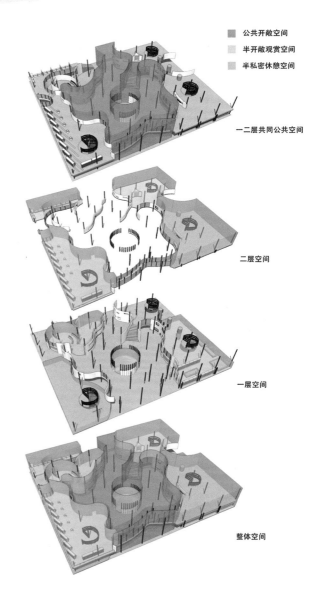

视线分析
Sight Analysis

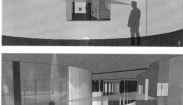

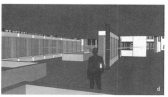

墙体的镂空让室内的空间变得更加丰富与无界限，玻璃与实墙的穿插变化关系，让视线变得更加生动。仰视、平视、俯视，室内空间其实是边界不那么明确，所以博物馆在塑造边界时也可以来创造这样模糊的边界，让边界更有意思。

The indoor space is made more ample and unbounded by the hollow-out wall. The sight becomes livelier as the conversion of the glass and solid wall. Looking up, straight and down, the indoor space actually does not have a clear boundary. So my museum can be created by these blurred boundaries that to make the boundary more interesting.

体验分析
Experience Analysis

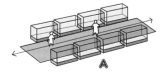
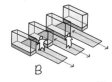
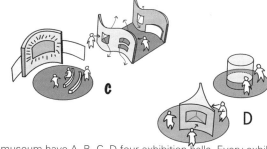

本身博物馆有 A、B、C、D 四个展馆，每个展馆各有自己体验方式，不管从视觉、听觉还有触觉与感觉。

A：展览方式以双排式，以最简单的形式进去这个博物馆，以最简单的形式去感受火器。B：袋入式的展示方式让游客意犹未尽。C：C 区可以观看影片，通过视觉感观去诠释火器，让游客更加了解，并且通过一步一景的感受去了解博物馆。D：D 区可以通过触摸的感受去体验火器的真实感。

The museum have A, B, C, D four exhibition halls. Every exhibition hall has its own experience style from visual sense, auditory sense and tactile sense.
A: The exhibition style of double-row is the simplest way to get into the museum, and also to understand the firearms. B: The display style of sack-in has not given full expression to the visitors' views. C: People can watch films at the C Zone, understanding the firearms by a different way that leads the visitors comprehend more and feel more about the museum. D: It is possible for visitors to feel the sense of reality of the firearms by touching.

节点效果图
Node Effect Drawing

学生感想
Students' Thoughts

南京作为一座历史古城，带给我们的历史记忆有太多太多。追溯时代的本质，从四大发明之一的火药到火器的历史进程，再到伴随战争的火器，到金陵机器制造局。方案否定了好多次，从空间出发，我总希望能将5.77m到9.17m的A1垂直空间做好，但是其实要保证流线与人流空间的通畅性，势必要浪费一些空间。所以，我选了火焰的概念，以一种自由的建筑空间姿态，是一个精神中心，然后去想象到底什么功能有这样作用，围绕着这个精神中心展开的展览，也就是一种新的空间形式，打破了一个空间串联一个空间的方式。最终效果可能并不尽人意，但是希望通过这次尝试建立更新的空间形式。

Nanjing as an ancient city with a long history brings us too many historical memories. Dating back to the essence of the age, it's the process from the powder belong to the Four Great Invitations to firearm, then to the firearm for war, and finally to the Jinling Machinery Manufactory Bureau. The project was denied many times. It was started from space. I tried my best to make the A1 vertical space from 5.77m to 9.17m. However, some rooms must be wasted to ensure the patency between streamline and human flow. So I chose the concept: flame. It's a free-style building space pattern. Besides, it's also a spirit center, and then people will image what kind of function could be effective like this. The exhibition focus on the spirit center is a kind of new space pattern. It breaks the rule that connect the rooms one by one. Maybe the result will not be good enough; I hoped to try the new pattern of the space.

教师点评
Teachers' Comments

《离火》以人与火、火器的关系展开设计概念，设计者以中国传统文化为大背景，将中国火器历史与战争巧妙融合，生成四个火器展示空间。主题明确，方案完整，体现了设计者比较专业的设计素质。思维逻辑不局限于对建筑空间的设计，更从人的视线，心理感受角度去探索空间的表达与意义。设计表现清晰明了，表达中规中矩。最终方案呈现出一定的可操作性，但还需进一步梳理建筑逻辑，推敲一些细节，加强空间结构与环境特色的整合与氛围塑造，增加参观者停留互动的元素。

——吕勤智　黄炎

Li fire develops the design concept based on the relationship among human beings, fire and firearm. The design, regarding the Chinese traditional culture as the main background, has integrated the history of Chinese firearm and war to form four firearm display rooms. The theme is definite and the scheme is integrated, having reflected the professional design quality of the designer. The designer's thinking logic is not limited to the design of the architectural space, but explore the special expression and meaning from view and emotion of human. The design presentation is vague and the expression is proper. The final scheme represents certain operability, but it requires further ordering of logic and modification of the details that the integration of special structure and environment as well as the atmosphere shaping should be reinforced and the factors attracting the visitors' attention and interaction should be added.

——Lv Qinzhi　Huang Yan

CIID"室内设计 6+1" 2015(第三届)校企联合毕业设计
CIID "Interior Design 6+1" 2015(Third Scssion)University and Enterprise Joint in Graduation Design

工业遗产保护与景观设计
Industrial Heritage Protection and Landscape Design

工业遗产保护与景观设计 Industrial Heritage Protection and Landscape Design

CIID"室内设计6+1"2015(第三届)校企联合毕业设计
CIID "Interior Design 6+1" 2015(Third Scssion)University and Enterprise Joint in Graduation Design

二等奖
Second Prize

45° 光景
45° Scene

高　　校：	华南理工大学
College：	South China University of Technology
学　　生：	马曼·哈山　吴晓飞
Students：	Maman·Hashan　Wu Xiaofei
指导教师：	姜文艺　陈建华
Instructors：	Jiang Wenyi　Chen Jianhua

关联页：154-162

设计理念
Design Concept

结合园区调研的感触和对符号的研究，我们将项目的名称确定为"45°光景"。45°角源自于兵工厂和法西斯两个符号的差异：第一这45°的差异象征着杀与止杀、战与之战两种不同的态度；第二45°是看问题的不同角度，同一把兵器，它可以是杀戮的攻击武器，也可以是阻止杀戮的守卫之兵，出发点不同，所得的结果也不同；第三相对于水平和竖直，45°是不稳定的状态，它是一种微妙的共生平衡。

Combining with our feelings on research of parks and our studying on symbols, we determine the name of the project as 45 ° scene. The angle of 45 ° comes from a difference between two symbols: Arsenal and Fasces. First, the difference of said 45° represents two different attitudes of killing and stopping killing, fighting and stopping fighting. Second, the angle of 45 ° is to see things from a different perspective. The same weapon may be used as an attacked weapon for killing, or be used as a guard for stopping fighting; different starting points lead to different outcomes. Thirdly, the angle of 45° is an unstable state relative to horizontal and vertical. It is a delicate symbiosis balance.

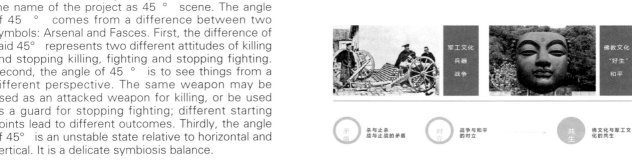

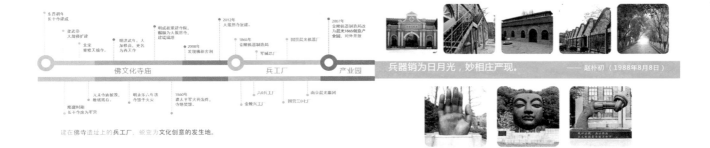

工业遗产保护与景观设计 Industrial Heritage Protection and Landscape Design

主动线规划分析
Active Line Planning

● 历史工业建筑
- 清朝
- 民国
- 50-90年代
- 游览主动线

● 商业服务网点
- 商业服务
- 游览主动线

● 主动线景观
- 水景观
- 绿化景观
- 展示台
- 休息亭
- 木质铺装
- 硬铺装广场
- 游览主步道

● 景观绿化
- 沿岸绿化
- 马家山绿化
- 新增绿化
- 游览主动线

● 展示空间
- 室外展示空间
- 室内展示空间

● 自行车服务
- 自行车流线
- 自行车服务点
- 游览主步道

● 历史遗存构件
- 工业遗存
- 佛文化遗存
- 铁轨
- 游览主动线

● 标识设立点
- 标识设立点
- 游览主动线

● 停留区域
- 硬铺装广场
- 游览主步道

"活"的博物馆
Active Line Planning

引入"活"的博物馆概念,将博物馆的展览划分为室内展览和室外展览两个部分。A1作为室内博物馆,主要展示武器制造,这是体现一个国家综合国力的硬实力,而园区内留存的遗址空间和人的活动则构成更加富有生气的室外博物馆。

室外的园区景观也是博物馆展示的一部分,由一条主动线将室内和室外的主要节点串联起来。

Introduce a living museum concept to divide exhibition of the museum into two parts: interior exhibition and outdoor exhibition. A1, as an interior museum, mainly exhibits weapon manufacturing. It's hard power of comprehensive national power of a country. The relic space and activities of people constitute a lively outdoor museum. The outdoor park scene is also a part of exhibition of the museum. The main nodes of the indoor and outdoor are connected in series by an active line.

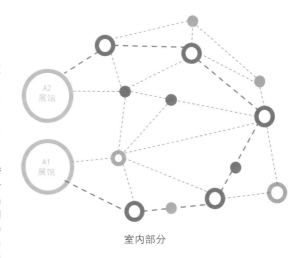

园区景观节点
Node of Park Landscape

场域:节点、动线形成一个强有力的场域,游客通过在场域内行走、驻足、感知到地块内的战争与菩提的矛盾共生的特质。

Field domain: the node and the active line form a powerful field domain where visitors walk, halt and perceive characteristics of a symbiotic contradiction between war and Bodhi.

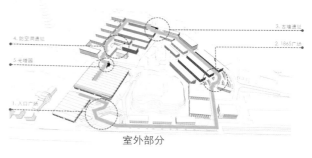

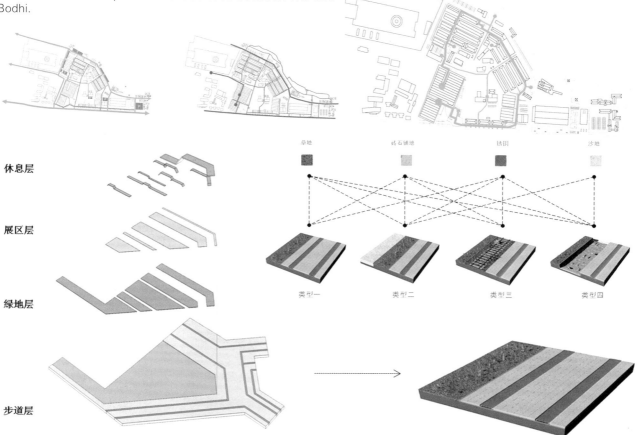

入口广场
Square Entrances

入口广场由民国时期和1949年建国以后的的两套网格组成，并用45°来巧妙地连接和融合两套网格。同时设计过程中对入口区域的交通、景观、地形分析应用于设计当中。

The square at the entrance is formed by two sets of grids at the period of the Republic of China and after the founding of the nation. The analysis of traffic, landscape and topography at the entrance are applied to design in the course of designing.

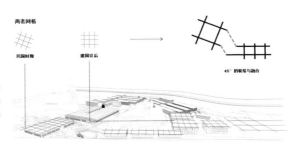

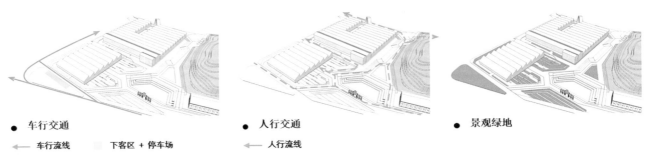

- 车行交通
- 人行交通
- 景观绿地

← 车行流线　　下客区 + 停车场　　← 人行流线

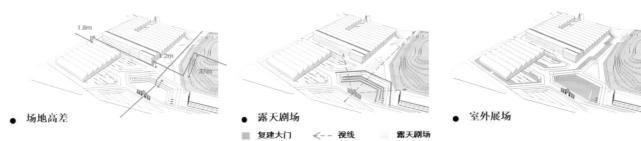

- 场地高差
- 露天剧场
- 室外展场

复建大门　　←-- 视线　　露天剧场

菩提园
Bodhi Park

工业遗产保护与景观设计 Industrial Heritage Protection and Landscape Design

光塔园
Guangta Park

光塔园位于园区游览动线的终点，光塔为游客提供一个在园区内登高望远的可能性。

Guangta Park is located at a terminal point of the sightseeing active line the park. It is a tower serving to provide possibilities for visitors look out over the tower in the parks.

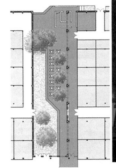

● 日照分析

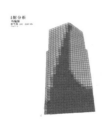
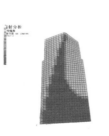

● 太阳能装置构造

利用分层钢板间的结构支撑，将太阳能蓄电池化整为零，分散在各个单独的小单元内。同一层的蓄电池通过钢板间的电线相互串联，共同为本层的LED灯带供电。

光电板设置在平板的平面上。通过资料调查，南京的太阳能利用最佳入射角为27.3度，光电板水平放置利用效率要高于数值放置。目的：设立的制高点即为光塔。

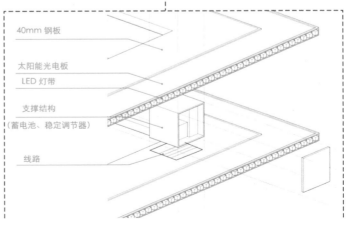

- 40mm 钢板
- 太阳能光电板
- LED 灯带
- 支撑结构（蓄电池、稳定调节器）
- 线路

学生感想
Students' Thoughts

随着毕业日子的到来，毕业设计也接近了尾声。经过十几周的奋战我的毕业设计终于完成了。在没有做毕业设计以前觉得毕业设计只是对这几年来所学知识的单纯总结，但是通过这次做毕业设计发现自己的看法有点太片面。毕业设计不仅是对前面所学知识的一种检验，而且也是对自己能力的一种再次提高。通过这次工业遗产建筑改造案例的毕业设计使我明白了自己原来知识还比较欠缺，自己要学习的东西还太多。总之我不仅温习了以前所学习的专业知识，还学到了很多关于保护历史建筑和改造设计知识，更深刻地体会到了同学之间、师生之间那份真挚的感情。

As the graduate date draws near, the graduate project is coming to the end. According to the effort for more than 10 weeks, my graduate project was finished at last. Before I did the project, I thought that the graduate project was only the summary of the knowledge I had learned. However, after doing the project, I found that my previous opinion was too one-sided. The graduate project is not only the test of the previous knowledge, but also the improvement of my own ability. I realize that I am lack of knowledge through the case of the industry relic reconstruction. I need to learn a lot. Anyway, I did not only review the previous knowledge I had learned, but also learn lots of knowledge about the historical protection and reconstruction. I have deeply understood the true friendships between teachers and students, and the true relationships among the classmates.

教师点评
Teachers' Comments

《45°-光景》作为兵工遗产主题的创意性设计方案，设计作者能够从历史的角度进行较为深刻的思考，提出较有新意的设计概念和设计切入点。设计方案的特点在于将建筑外部空间的遗址环境作为博物馆的组成部分，以景观设计的手法针对遗存的外部空间和历史遗迹，将园区主题与外部空间功能进行有机整合，使园区的整体环境具有统一的主题氛围，同时满足人们室外活动的需求。设计方案在概念生成、设计分析、主题立意、方案深化、成果呈现等方面的工作较为系统完整，表现出设计小组成员较突出的综合设计素质。

——吕勤智

45° Scene is a creative design project, and the theme is ordnance heritage. In the view of the history, designer could think deeply and put forward an innovative concept and a point of penetration. The characteristic of the project is that the relics outside the building are included as parts of the museum. To the outer space and historical relics, their functions are combined with the theme of the park closely according to the landscape design method. It makes the general environment has a unique theme atmosphere, meanwhile satisfied people's needs of outer space activities. The project is completed in many aspects, such as: concept generation, design analysis, theme meaning, project deepening, and result showing and so on. It shows the overall quality of each design group member.

——Lv Qinzhi

专家点评
Experts' Comments

设计者对原有的厂区有较好的保护和规划意识，以改善园区的办公环境为出发点，对植物安排和道路景观进行了精心的规划，通过设计提升了旧厂区的整体环境景观质量水平。对厂区入口交通规划设计手法大胆，对人车分流的规划上有一定的想法，但在投资规模与使用效益上还要做一定的考量。

——林学明

The designer has a good awareness of the original factory's protection and programming. In order to improve the office environment in the park, the plant arrangement and road landscape were carefully programmed. According to the project, the general environment of the old factory and landscape's quality are promoted. The method of the traffic entrance program is adventurous. There is an idea about the diversion of people and cars. However, the scale of investment and the benefit should be carefully considered.

——Lin Xueming

CIID "室内设计 6+1" 2015（第三届）校企联合毕业设计
CIID "Interior Design 6+1" 2015(Third Scssion)University and Enterprise Joint in Graduation Design

二等奖
Second Prize

止戈为林
Stop Fighting for Forest

高　　校：	**哈尔滨工业大学**
College :	Harbin Institute of Technology
学　　生：	**贾思修　乔红　韩思宇　王岩　王子轩**
Students :	Jia Sixiu　Qiao Hong　Han Siyu　Wang Yan　Wang Zixuan
指导教师：	**马辉　刘杰**
Instructors:	Ma hui　Liu Jie

关联页：90-99、162-169

空间营造
Space Construction

以人的行为模式确定空间性质及要素，利用景观体系营造空间。
To determine space properties and elements in human behavior patterns, construct space by utilizing landscape system.

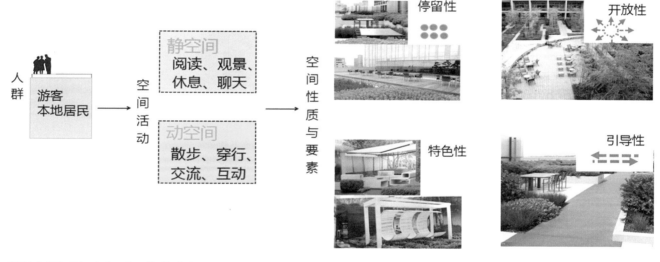

场地景观形态生成分析图
Analysis Diagram of Site Landscape Morphology Generation

从 A1、A2 馆立面锯齿形态中提取出单个锯齿形态。通过对此形态的延伸、叠加形成不同状态的相同样式的元素。并根据场地轴线的趋势进行变化组合，结合场地功能需求，形成此次方案设计的场地形式。

Extract individual saw tooth morphology from facade saw tooth morphology of pavilions A1 and A2. By extending and overlapping this morphology to form elements of different states and the same pattern. Make changes and combinations according to a tendency of site axis and form the site form of this scheme design by combining with requirements of the site function.

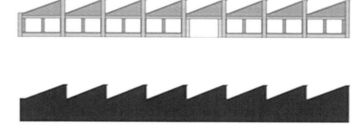

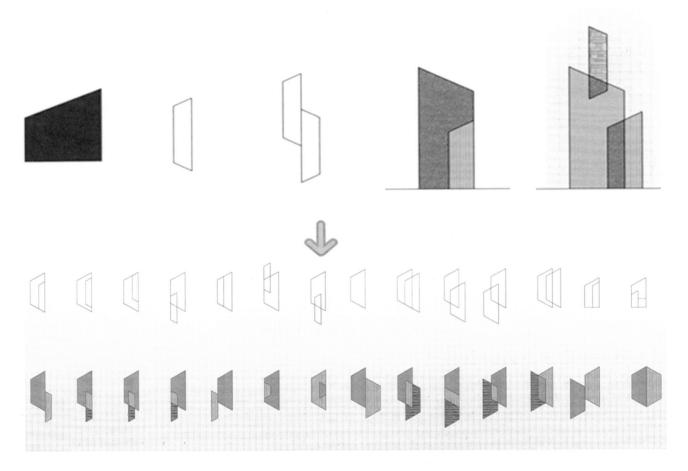

设计说明
Design Explanation

通过对场地周边环境的规划和发展目标的研究，本设计根据游览路线的设定和研究，并结合基地环境和周边环境的现状和历史，对区域进行景观改造设计过程，尤其注重历史建筑保护，力图打造一个满足活动、休憩、娱乐等现状的场所。

Through planning surrounding environment of the site and studying goal of development, the design aims at landscape reconstruction of the areas according to setting and studying of a touring route combining with current situation and history of base environment and surrounding environment, especially focusing on historical buildings preservation, to create a place for satisfying activities, rest and entertainment.

Industrial Heritage Protection and Landscape Design　工业遗产保护与景观设计

Reviving in the Dusk to Dawn
基地分析
Base Analysis

基地位置
Base Position

区域示意
Base Position

区位关系
Base Position

功能分布
Base Position

交通状况
Base position

止戈为林 工业遗产下的精神与生态的再生

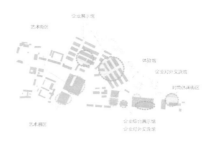

　　南京晨光1865创意产业园所在地为李鸿章于1865年兴建的金陵制造局旧址，如同一座近代中国工业博物馆，记录着中国民族工业发展的历史轨迹。

　　Nanjing Chenguang 1865 Creative Industry Park is located at a former site of Jinling Manufacturing Bureau constructed by Hongzhang Li on 1865. It is like a modern Chinese Industry Museum, recording historical track of development of Chinese national industry.

　　秦淮区因秦淮河贯穿全境而得名，是古都金陵的起源，秦淮文化是金陵文化的精华，有"江南锦绣之邦，金陵风雅之薮"的美称。

　　Qinhuai district is named from Qinhuai River flowing through the whole district. It is an origin of the ancient city Jinling.Qinhuai culture is an essence of Jinling culture and the Qinhuai district wins the laudatory title of a splendid city of Jiangnan district, an artistic place.

　　秦淮区是南京市的中心城区，国家东部地区重要的金融商务中心，华东地区的商贸、信息、文化、旅游中心，现代化国际性人文绿都核心区之一。

　　Qinhuai district is a central city area of Nanjing, an important financial business center of eastern region, a commercial and trade center, an information center, a culture center and a tour center, and one of core areas of the modern international humanity green city.

　　秦淮区周边主要分布着学校、住宅、办公、商业、剧场、医院以及城市绿地等功能用地。这对我们对建筑功能的定义具有重大意义。

　　Qinhuai district is surrounded by functional lands such as school, house, office, business, theatre, hospital and city green land. It has a significant meaning to the definition of building function.

　　交通方面，秦淮区现有公路、铁路、航空以及地铁。现地铁运营中的有南京地铁1号线、2号线、3号线。

　　Transportation: at the present, Qinhuai district has highway, railway, aviation and subway. The current metro operation includes Nanjing metro line 1, line 2 and line 3.

气候分析
Climatic Analysis

南京风玫瑰图
Wind-rose Diagram of Nanjing

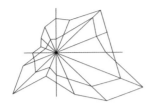

南京盛行东南风、西北风；夏季主导风向为东南风、东风；冬季主导风向为东北风、东风。

南京月温度分析图
Nanjing Month Temperature Analysis Chart

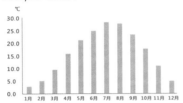

南京属亚热带季风气候，冬冷夏热、四季分明。南京市年平均气温 15.9℃，年极端气温最高 39.7°C，最低 –13.1°C。

南京月平均蒸发量分析图
Nanjing Month Average Evaporation Capacity Analysis Chart

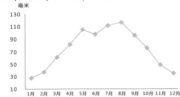

南京月平均蒸发 8 月最多，占全年的 12.8%；7 月次之；1 月最少。1~5 月逐渐增多，6 月略有回落，7、8 月为全年最高值，9 月开始减少。

南京日照分析图
Nanjing Sunlight Analysis Chart

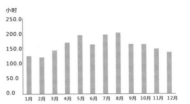

南京光照充足，每月日照时长相近，夏季日照时长多于其它季节，春季日照则相对较少。

南京月降水量分析图
Nanjing Month Amount of Precipitation Analysis Chart

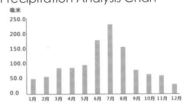

南京属亚热带季风气候，冬冷夏热、四季分明。南京市年平均气温 15.9℃，年极端气温最高 39.7°C，最低 –13.1°C。

SWOT 分析
SWOT Analysis

	积极	消极
	优势因素（Strengths）	劣势因素（Weaknesses）
内在	1.位于南京市城市副中心的旧厂区，具有浓厚的商业和文化氛围。 2.北临秦淮河，自然景观丰富，吸引人群汇集。 3.街区内保留了不同历史时期的建筑，具有重要的研究价值。	1.街区联系不紧密，缺乏有效的设计方法。 2.历史工业文化未得到有效传播。
	机会因素（Opportunities）	挑战因素（Threats）
外在	1.历史街区前所未有的重视。 2.城市的品牌需求 3.周边旅游产业密集	正确定位：如何根据现状条件对地块进行正确定位，影响到地块未来的发展方向。 保护与开发的矛盾：如何使历史街区在新世纪具有持久的生命力是这个地块对我们最大的挑战。

总平面图分析
General Layout Analysis

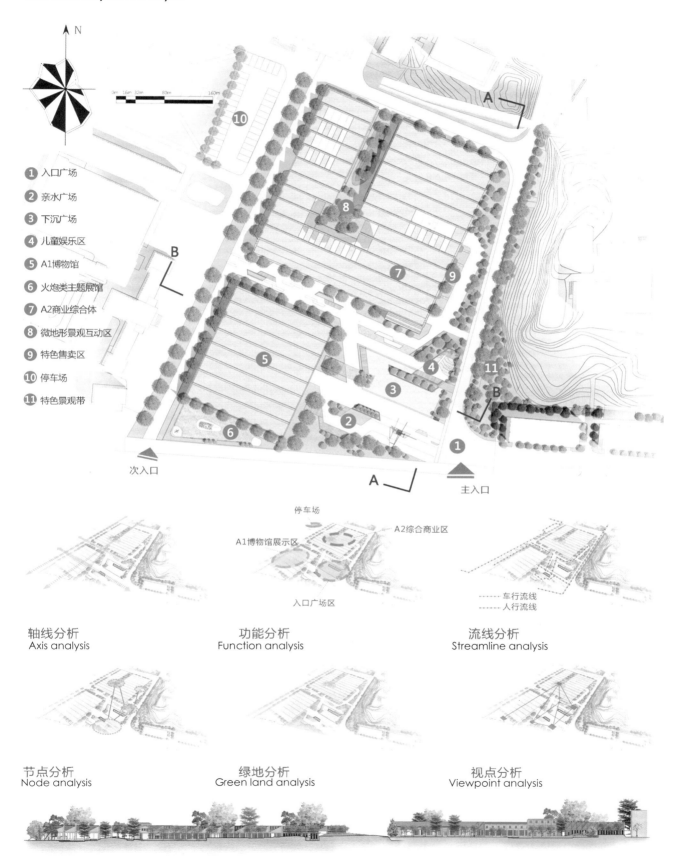

① 入口广场
② 亲水广场
③ 下沉广场
④ 儿童娱乐区
⑤ A1博物馆
⑥ 火炮类主题展馆
⑦ A2商业综合体
⑧ 微地形景观互动区
⑨ 特色售卖区
⑩ 停车场
⑪ 特色景观带

轴线分析
Axis analysis

功能分析
Function analysis

流线分析
Streamline analysis

节点分析
Node analysis

绿地分析
Green land analysis

视点分析
Viewpoint analysis

工业遗产保护与景观设计 Industrial Heritage Protection and Landscape Design

	地被及小灌木			大灌木		小乔木					大乔木				人文景观			自然景观				习性特征 HABITS AND
	金丝桃	细叶麦冬	素馨	杜鹃	红花继木	栾树	唐菖蒲	珊瑚树	鹅掌楸	水杉	悬铃木	香樟	雪松	国槐	连翘	桂花	柳树	羽衣甘蓝	一串红	月季	玉兰	
耐寒	○	●	○	●	○	●	○	●	○	●	○	●	○	●	○	●	○	●	○	●	○	
耐盐碱	○	●	○	●	○	●	○	●	○	●	○	●	○	●	○	●	○	●	○	●	○	
耐水湿	○	●	○	●	○	●	○	●	○	●	○	●	○	●	○	●	○	●	○	●	○	
抗病虫害	○	●	○	●	○	●	○	●	○	●	○	●	○	●	○	●	○	●	○	●	○	
喜光	○	●	○	●	○	●	○	●	○	●	○	●	○	●	○	●	○	●	○	●	○	
耐阴	○	●	○	●	○	●	○	●	○	●	○	●	○	●	○	●	○	●	○	●	○	适应环境 ENVIRONMENT
低洼地																						
沙质土壤																						
林下																						
路旁																						形态 MORPHOLOGY
香味	●	○	●	○	●	○	●	○	●	○	●	○	●	○	●	○	●	○	●	○	●	
观茎	●	○	●	○	●	○	●	○	●	○	●	○	●	○	●	○	●	○	●	○	●	
观花	●	○	●	○	●	○	●	○	●	○	●	○	●	○	●	○	●	○	●	○	●	
观叶	○	●	○	●	○	●	○	●	○	●	○	●	○	●	○	●	○	●	○	●	○	

Industrial Heritage Protection and Landscape Design　工业遗产保护与景观设计

CIID"室内设计 6+1" 2015（第三届）校企联合毕业设计
CIID "Interior Design 6+1" 2015(Third Scssion)University and Enterprise Joint in Graduation Design

三等奖
Third Prize

生长和碰撞
Growth and Crash

高　　校：	哈尔滨工业大学
College:	Harbin Institute of Technology
学　　生：	金喆　伏祥　何佳佳　张泽宇
Students:	Jin Zhe　Fu Xiang　He Jiajia　Zhang Zeyu
指导教师：	马辉　刘杰　周立君
Instructors:	Ma hui　Liu Jie　Zhou Lijun

关联页：100-105、169-173

金喆 Jin Zhe　　伏祥 Fu Xiang　　何佳佳 He Jiajia　　张泽宇 Zhang Zeyu

生长与碰撞——南京1865产业园景观设计
Growing and Crashing—Nanjing 1865 Industrial Park Landscape Design

总平面图 General layout

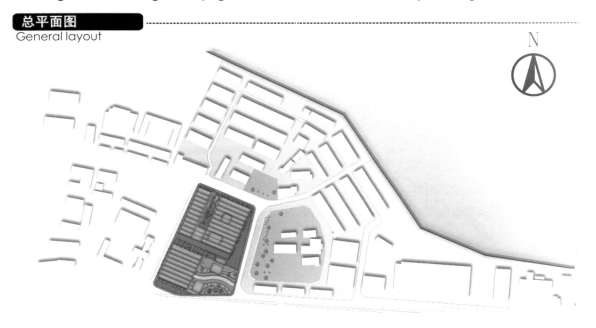

在我们对相似的一些项目进行调研之后，我们发现，在国内的类似项目中，多少出现了文化产业的发展向纯商业转移的趋势。这种情况在业界普遍被认为是一种不可逆转的趋势。在本次设计中，我们试图去改变这样的现状。我们在设计中，利用串联A1,A2两馆的步道系统，将商业部分和文化部分分隔开，商业部分更多的利用步行流线，而艺术家工作区域更多的使用车行流线。我们用廊道和绿化弱化了工业建筑的"硬度"，我们对于场地的定位是一个以近代军事工业博物馆为主题的综合体，我们在室外部分也准备了室外展示的室外展厅，融入我们观展流线的一部分。

After research of some similar projects, we find tendency of transfer from culture industry to pure commerce in many similar projects in China. This kind of situation is widely considered to be an irreversible trend in the industry. In this design, we try to change such a current situation. As such, we separate the commercial part from the culture part by means of footpath systems of A1 and A2 pavilions in series. We utilize walking streamline for the commercial part more and driving streamline for the area where artists work. We weaken "hardness" of industrial buildings by means of gallery and greening. The site is positioned at a complex of modern military and industry museum. We also prepare the outdoor exhibition hall to be a part of our exhibition-viewing streamline.

Industrial Heritage Protection and Landscape Design 工业遗产保护与景观设计

生长与碰撞——南京 1865 产业园景观设计
Growing and Crashing—Nanjing 1865 Industrial Park Landscape Design

步行街部分
Pedestrian Street Part

在本次的设计中，我们保留了原有建筑群中的大多数树木。在建筑中，有街墙一说，产生道路的因素是建筑本身而不是由道路而产生了建筑。A1、A2 所夹着的道路便是如此。除了建筑之外，数目也构成了这一径向通道的一部分。我们用链接 A1、A2 的构造桥将其破开，使得其接到的元素更加丰富。

在本次设计中，我们遇到的问题之一，就是如何调和 A1 馆的博物馆和 A2 商业建筑在功能上的冲突。这也是我们最早发生的碰撞。当然，我们用生长的方法解决了这个问题。A2 馆的入口，就用 A1 馆的尾声作为开始。两者的链接，包含我们对于近代中国历史的印象：在曲折中前行，在挑战中不断强盛。

In the design, we reserve most trees on the original buildings. There is a saying about street wall in the building, which means that it is the building itself that generates the road. So is the road between A1 and A2. Besides, the trees also constitute a part of this radial passageway. We break it open by utilizing a construction bridge that links A1 to A2 such that the elements linked thereto are more plentiful.

One of the problems we encountered in the design is how to reconcile a conflict on function between a museum of A1 pavilion and A2 commercial building. This happens to our earliest crash. But certainly, we solve this problem through the method of growing. The end of A1 pavilion is used as a start of an entrance of A2 pavilion. By linking those two, we present our impression to modern Chinese history: to move ahead in twists and turns, and never stop to be powerful and prosperous in challenges.

生长与碰撞——南京 1865 产业园景观设计
Growing and Crashing—Nanjing 1865 Industrial Park Landscape Design

休憩
Rest

在靠近主干道的区域，我们设置为针对市民的活动场所，设置休闲座椅等，方便附近的居民和园区内的工作人员。

Close to the main line, we set up an activity place for citizens with leisure chairs to facilitate the nearby residents and staff in the parks.

展示
Demonstration

A1 前的广场同样起到室外展陈的作用。军事类博物馆通常利用砂石地面作为大型室外展品的陈列区域。

The square in front of A1 also serves to the outdoor exhibition. The military-based museum generally utilizes dinas ground as display areas of large-scale outdoor exhibits.

生长与碰撞——南京 1865 产业园景观设计
Growing and Crashing—Nanjing 1865 Industrial Park Landscape Design

步道
Footpath

A1、A2 之间的部分同时体现"道"的功能和"庭"的功能。因此，我们增加了植被，同时增加了场地休憩的功能。因为廊道的存在，人流被分流，我们希望步道区域适合那些悠闲的旅行者，而廊道适合追求效率的工作人员。

The part between A1 and A2 embodies functions of "road" and "courtyard" at the same time. As such, we increase the part of vegetation and the function of site rest. Attribute to the presence of gallery, people are diverted. We hope that the footpath is suitable for leisure travelers and the gallery is suitable for efficiency-pursuing staff.

廊道
Gallery

廊道连通了工作区和 A2 的正门，同时具有较好的并行性能。位于上方的廊道同时可以俯瞰庭院、步道。

The gallery connects working area to a main entrance of A2 and at the same time, has a better parallel performance. The people standing on the above gallery can overlook the courtyard and footpath.

灯光
Light

根据昼夜适用人群的不同，我们在灯光的使用上考虑了灯光的导向性，在白天利用自然光强调庭院空间，晚上利用照明强调廊道的功能性。

According to day and night suitable for different people, we consider orientation of the light and utilize daylight to emphasize the courtyard space by day and illumination to emphasize functionality of the gallery by night.

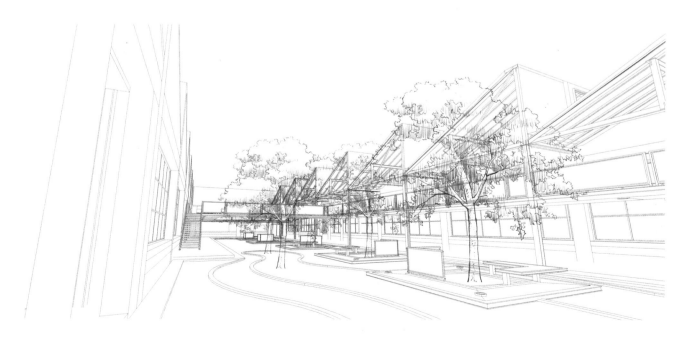

工业遗产保护与景观设计 Industrial Heritage Protection and Landscape Design

CIID "室内设计 6+1" 2015(第三届)校企联合毕业设计
CIID "Interior Design 6+1" 2015(Third Scssion)University and Enterprise Joint in Graduation Design

三等奖
Third Prize

融·生
Convergence • Generation

高　　校：	浙江工业大学
College:	Zhejiang University of Technology
学　　生：	张泽浩
Students:	Zhang Zehao
指导教师：	吕勤智　黄焱
Instructors:	Lv Qinzhi　Huang Yan

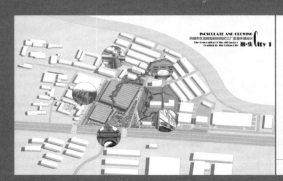

关联页：174-181

张泽浩
Zhang Zehao

设计说明
Design Explanation

随着城市发展，废弃厂区成为城市发展的消极因素，希望用一种巧妙的方式，在不破坏历史街区的情况下，为厂区营造更多的公共开放性空间，满足市民文娱体等活动可能，让封闭厂区重新融入到城市中。

晨光1865园区是历史街区、兵工遗产、文物建筑，园区内建筑分布散乱，路网密集，道路等级低。受文物建筑保护和历史街区的限制性因素，不能大拆大建，因此提出了一个景观的设想来解决存在的问题，作为一个独立的景观系统，来探讨景观结构的可能。

这个新的空间模式将历史街区的交通压力排解到地下和地上。地上是以观光功能为主的景观廊道，这样不仅缓解了历史街区的交通流量压力，也为游客提供了一个游憩观光的空间。地下是以公共开放性空间为主，为园区提供了必要的刚性需求以及满足市民文娱活动的可能。这一构想无疑将会改变人们习以为常的城市形象，重新唤醒历史街区的活力，提升城市的气质，推动城市的发展。

With the development of the city, the abandoned factories become negative factors of development of the city. We hope to create more public open spaces for the factories without destroying historical blocks in an ingenious way to satisfy requirements of citizen's recreational activities so that the closed factories can be re-integrated into the city.

Chenguang 1865 parks belong to historical blocks, ordnance factory heritages, and cultural relic buildings. In the parks, the buildings are in sprawling ill distribution, the road network is dense and the road grade is low. Under the influence of restrictive factors of cultural relic buildings protection and historical blocks, the parks do not allow large-scale demolishing and reconstruction; in view of this, we put forward an idea of landscape to solve the current problem and take it as an independent landscape theory to discuss possibilities of a landscape structure.

This new spatial pattern will relieve traffic pressure of the historical blocks to both over and underground. The above ground mainly includes landscape corridors that are sightseeing-oriented. This will not only ease traffic pressure of the historical blocks, but also provide a rest space for tourists to go sightseeing. The underground mainly includes public open spaces, serving to provide necessary rigid demands for the parks and satisfy requirements of the citizen's recreational activities. This vision will undoubtedly change city image that people are accustomed to, reawaken vitality of the historical blocks, enhance temperament of the city and promote development of the city.

现状问题的提出
Presentation of Modern Problems

交通问题 Transportation Problems	空间结构 Spatial Structure	生态优化 Ecological Optimization	周边渗透 Surrounding Penetration

狭窄、闭塞的道路与大量的车流、人流间的矛盾。 Contradiction between the narrow and blocked roads and the large influx of vehicles and people.	建筑高密度，空间系统凌乱，缺乏大量公共空间。 Buildings of high density, messy space system, lacking of a large number of public spaces.	园区绿地分布不均，相对独立并缺少联系。 The park greens are ill-distributed, independent from one another and lacking of connections.	厂区各部分之间缺少联系衔接，彼此无法渗透。 All parts of the factories are lacking of contact cohesion and penetration.
提出立体交通的设想，将人车分流减轻地面交通压力。 Contradiction between the narrow and blocked roads and the large influx of vehicles and people.	通过对立体空间的设计，增加大量公共空间。 Through the design of stereo space to increase a lot of public space	扩充、延伸绿地面积，提出生态立体空间的理念。 Expand and extend the green area, and put forward a theory of ecological stereo space	通过立体空间的模式，把厂区的各部分空间联系起来。 Connect all parts of the factories to one another via a stereo space pattern.

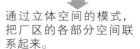

工业遗产保护与景观设计　Industrial Heritage Protection and Landscape Design

提出新的空间模式
Put Forward a New Space Pattern

空中观光系统：
观光游览
Air tour system: sightseeing

架空层景网系统：
各种社交行为
Stilt floor landscape network system: various social behaviors

地面交通系统：
混合功能主导
Ground transportation system: mixed function dominated

地下空间系统：
功能性主导
Underground space system: functionality dominated

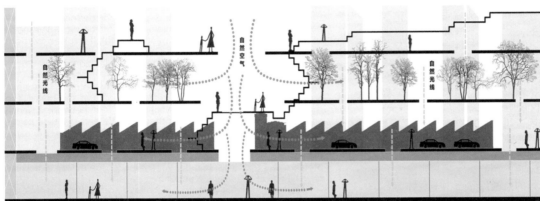

生态的空间模式
- 树叶空间：景网系统
- 树枝空间：景观廊道
- 树干空间：地面交通系统
- 树根空间：地下公共空间

无障碍步行系统
Barrier-free Walking System

　　在地面空间上再架设一个空间，加入功能载体，把室外当室内用，空间层次被模糊了。组团之间的交流从地面空间解放出来。

　　Setting up a new space on the ground space and joining function carrier, the outdoor space used as indoor, and then spatial hierarchy is blurred. The communication between the groups is liberated from the ground space.

方案生成
Scheme Generation

原生 Protogenesis　　　　**新生** Renascence　　　　**融生** Convergent Generation

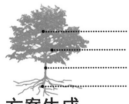

文化
佛文化与工业文明

肌理
密集的古建筑群与街区

文化
历史文脉的延续与现代文化的注入

产业
增加商服调整产业结构

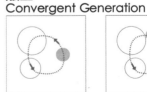

一超多强的文化氛围　　多产业协调发展

生活
厂区生活

建筑
历史建筑+工业厂房

空间
增加廊道与各个空间的联系

活动
增加公共开放性空间与外部活动

内外各层空间的交融　　公共私密空间的和谐

Industrial Heritage Protection and Landscape Design 工业遗产保护与景观设计

方案设计
Scheme Design

拆除周边质量较差、对历史建筑有压迫感的建筑和树。
Demolish the surrounding buildings and remove trees, which has a poor quality and a sense of pressure to the historical buildings.

延续历史建筑的主体感和轴线关系。
Extend the body of the historical building and axis relationship.

增加绿地的延伸，让绿地与开放性空间形成缓冲。
Extend grassy area to form a buffer between the grassy area and the open space.

通过节点的深化，用观景廊道将其串联形成联系。
By deepening of the joint, connect them in series with landscape corridors to form the connection.

方案分析
Scheme Analysis

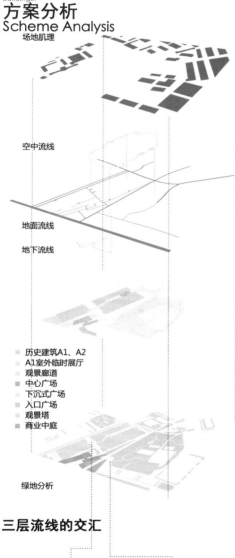

建筑体量对比
Building Size Comparison

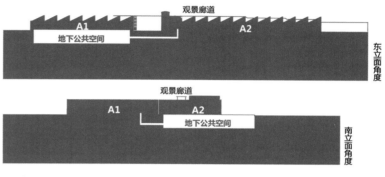

架空与下沉双线发展

以点带面的延续
Continuation of Point-to-area

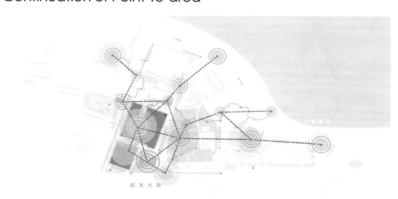

三层流线的交汇

景观廊道分布
Landscape Corridor Distribution

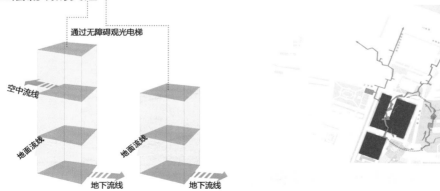

园区立面图
Elevation Diagram of the Parks

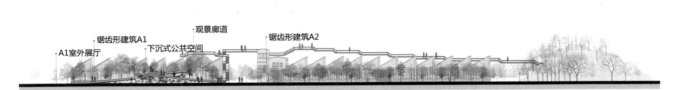

节点设计：中心广场
Joint Design: Central Plaza

绿色生长策略
Green Growth Strategy

平面上的绿色生长
Green Growth on the Plane

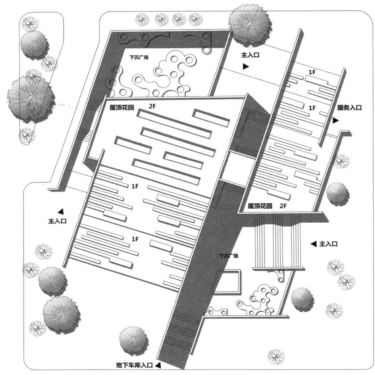

剖面上的绿色生长
Green Growth on the Section

立面上的绿色生长
Green Growth on the Vertical Face

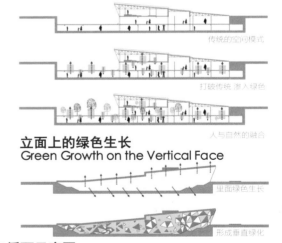

水循环示意图
Schematic Diagram of Water Cycle

空气循环示意图
Schematic Diagram of Air Circulation

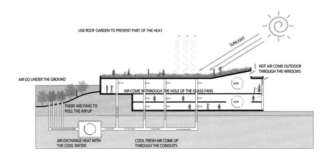

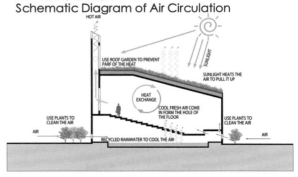

Industrial Heritage Protection and Landscape Design　工业遗产保护与景观设计

首层平面图
Plan of Ground Floor

地下一层平面图
Plan of the First Floor Underground

地下二层平面图
Plan of the Second Floor Underground

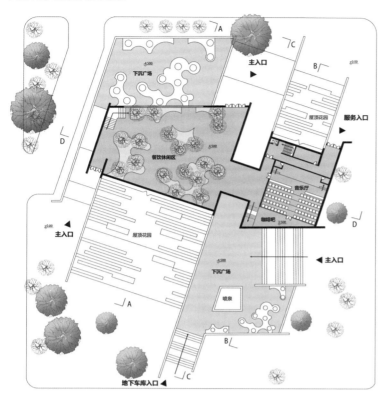

剖面图
Sectional Drawing

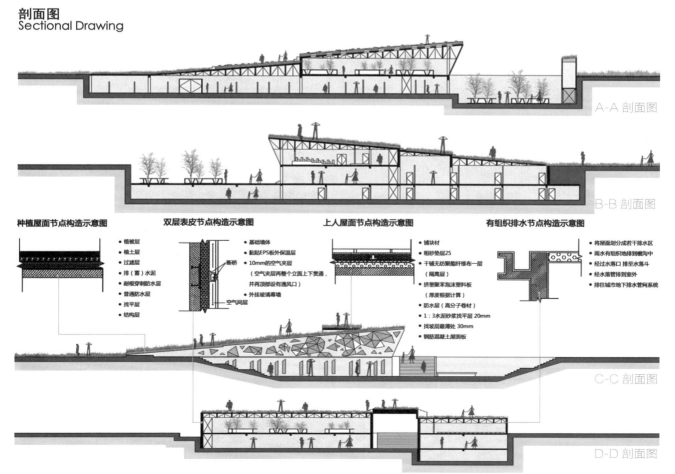

179

效果图
Design Sketches

学生感想
Students' Thoughts

此次毕业设计为我们提供了一个多校交流的平台，通过与其他院校师生的交流，感受不同的思维方式与表现形式，收益良多。

虽然大家来自不同的学校，但通过这次毕业设计，我们从陌生变得熟悉，从竞争对手变成了朋友，这样一份友谊难能可贵。

比起最后的结果，我更享受的是学习、历练的过程，我享受这个过程带给我的成长。所以整个设计过程对我来说，弥足珍贵。

回想起整个毕业设计过程，一路磕磕绊绊，有酸有甜，方案阶段在不停的否定与自我否定中成长，真的很庆幸自己能在大学本科的最后阶段，有这么一个锻炼自己的机会和平台。

This design project provided us a multi-college communication platform. According to the communication with students from other colleges, I felt different thought and expression, and benefited a lot.

Although we were from different colleges, our relationship was getting closer and closer according to the project. We turned the rivals to the friends. This kind of friendship is precious.

Compared with the final result, I prefer the process of the study and experience. I enjoyed my growth through the process. So each process of the design is precious for me.

Remind of the whole process of the design, it's very hard. The process was filled with happiness and sorrow. During the project stage, we grew up through the continuous negation and self-negation. It's my fortune that I could have a chance to improve myself at the end of the stage in the university.

教师点评
Teachers' Comments

本组设计着力于从架空与下沉双线发展，让游客能够最大限度感受历史建筑和历史文化的魅力，选择了将建筑埋于地下和被架起，而将地面空间充分保留，形成一个四方八达的城市广场。丰富的空间场地设计在满足功能定位的基础上，很好地保护和利用了工业建筑的独特属性，较为合理地重塑人性化的空间环境。其中值得商榷的是，架空形成一条来回摆动的折线，力图通过左顾右盼的折线形成曲桥，但是较为均质的景观节点似乎很难达到延长风景线、扩大景观画面的效果。

——刘杰

The group's design focus on the two ways development: the impending one and the underground one. It lets the tourists feel the charm of the historical buildings and ancient culture in maximum degree. They tried to make the buildings be built on stilts, or be buried underground, and then reserved the floor space to form an all direction extending city square. The square with abundant space does not only satisfy the functional orientation, but also protect and use the unique property of the industry building. The humanized space environment is rebuilt reasonably. On the other side, there is something to discuss. After the building is built on stilts, a pendular fold line is created to form the curved bridge. However, the homogeneous landscape node may not reach the extended scenery line and expend the landscape frame.

——Liu Jie

专家点评
Experts' Comments

方案试图从景观和交通系统的角度对厂区及周边的环境进行整体设计，并提出以下沉广场和架空绿化形成步行连廊系统的设计方案。想法大胆，手法亦比较活跃，其中构思的景观塔也很有想象力。不可否认，这是一个很有创意的方案。遗憾的是，其呈现的效果对原旧厂房风貌有所破坏，精心设计营造的空间环境实用性略显不足，且其实际空间与所构想的功能所需空间不尽相符。这也许是创作者在出发点上考虑欠周，但仍不失为一个富有想象力的方案。

——冼剑雄

In the view of the landscape and traffic system, the designer tried to made a general project of the factory and its surrounding environment, and put forward a plan that formed a walking gallery according to the sunken square and overhead greening. The thought is adventurous, and its method is active. Especially the thinking of view tower is imaginative. However, it's a pity that the effect breaks the original view of the old factory. The well-designed space environment is slightly impractical. In fact, the space area could not meet the requirement for the imagined function. Maybe the designer's thought was not considerable in the beginning, but it's really an imaginative project.

——Xian Jianxiong

工业遗产保护与景观设计 Industrial Heritage Protection and Landscape Design

CIID "室内设计 6+1" 2015（第三届）校企联合毕业设计
CIID "Interior Design 6+1" 2015(Third Scssion)University and Enterprise Joint in Graduation Design

佳作奖
Excellence Prize

穿越金陵 2065
Passing Through Jinling 2065

高 校：	浙江工业大学
College:	Zhejiang University of Technology
学 生：	孙莹
Students:	Sun Ying
指导教师：	吕勤智　黄焱
Instructors:	Lv Qinzhi　Huang Yan

关联页：182-187

Industrial Heritage Protection and Landscape Design 工业遗产保护与景观设计

孙莹
Sun Ying

区位分析
Location Analysis

园区位于南京市中心地区南侧，处于历史文化氛围十分浓郁的秦淮区，园区的开放性低，与居民区互动性低；东西侧门入口小，辨识度低。人车混行，园区缺少活动场地以及能引起聚集行为的场地。

The park is located in south of central region of Nanjing, situating in Qinhuai district where the historic culture atmosphere is very rich. The openness of the park is low and the park has a low interactivity with residential areas. The entrances of east and west side doors are small and thereby have a low identification. At there, people and vehicles move in mix. The park is short of activity place and the place that can cause gathering behavior.

设计愿景
Design Vision

设计对象
Design Objects

三类人群每日户外停留时间和流量分析
Analysis of outdoor dwell time and flow of three categories of people per day

总结：人群户外停留时间少，缺少互动性和自由创作空间。周边居民区以及规划建设的商业综合体和办公区域，具有扎实的人气基础。

Conclusion: the people have a little outdoor dwell time and do not have the space for interactivity and free creation. The surrounding residential areas are numerous, so that the commercial complex and office areas to be planned and constructed would have a strong popularity.

园区现状分析
Analysis of Current Situation of The Park

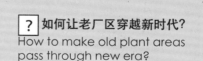

人行流线分析
Streamline analysis of people walking

车行流线分析
Streamline analysis of driving

景观空间分析
Landscape space analysis

? 如何让老厂区穿越新时代？
How to make old plant areas pass through new era?

概念演绎
Concept Deduction

 →  →  → 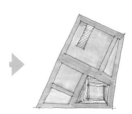 →

设计理念
Design Concept

穿越金陵2065，实现人与建筑的共生，过去、当下与未来的共生，一个既属于历史建筑又属于当代人的景观空间。也是穿越两个时空的纵向空间与服务各方需求的横向空间，以当代的设计手法抽象演绎过去式的元素，以当代的设计元素探索式表现未来的概念。做专属1865的景观空间，是不可以被复制、不可以被移动的。

Passing through the Jinling 2065 to achieve coexistence of human and buildings and the coexistence of the past, now and future to create a landscape space belonging to not only the historic buildings but also the modern people. It is a longitudinal space passing through two time spaces and a transverse space meeting requirements of each side. It is an element of abstract deduction of the past tense in the contemporary design skill and a concept expressing future by means of exploring on the basis of the current design element. It cannot be duplicated and moved to make the landscape space for the exclusive 1865.

剖面图
Design Concept

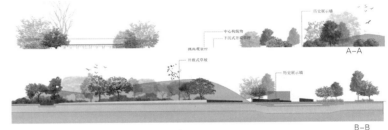

总平面图
General Layout

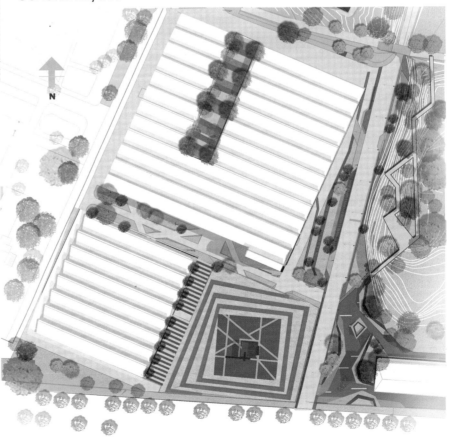

分析图
Analysis Chart

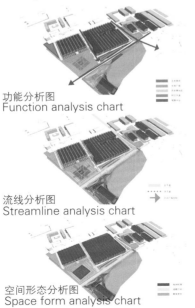

功能分析图
Function analysis chart

流线分析图
Streamline analysis chart

空间形态分析图
Space form analysis chart

中央草坪设计意向
Central Lawn Design Intension

形式体现：结合原场地周边现状与道路走向，整理出四边形中央草坪，与A1、A2建筑形态在形式上相互协调，形成有序的地块机理，遵循潜在逻辑关系，在变化中有统一。

Form embodying: combining with surrounding current situation and trend of road, we set up a quadrangle central lawn to inter-coordinate with A1 and A2 building forms on the form so as to form a well-organized block mechanism by following the guide that the potential logic relationship has unification in the change.

主题体现：感受园区历史气息，穿越金陵1865~2065，真正的尊重历史是（保护旧物原貌并发掘其在当代存在的意义发挥它的价值并延续使用它）使历史旧物原地原貌重现。

Theme embodying: to feel historical breath of the park, to pass through Jinling 1865-2065; to respect the history in real (to protect the original appearance of the ancestors' heritage and find the meaning of its presence at the present times, to play its value and continuously use it) is to reproduce historical ancestors' heritage in situ on the original appearance.

中央草坪效果图
Design Sketch of Central Lawn

 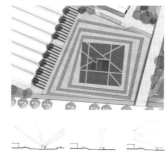 

地块的下沉感给人聚拢之意，下沉楼梯用印有时代象征性纹样的旋转楼梯，犹如行走在园区历史的时空中，越向中间行走越加强烈的历史感悟感，走到草坪中央触摸大型构筑物达到历史感悟的高潮。此地块可用来放露天电影、音乐旱喷、夜晚灯带导向指示，增加园区人们的户外活动以及停留时间。达到在博物馆前聚拢人气，提高博物馆建筑的招示性目的。

The sense of sinking of the plot shows a meaning of assembling. The sunken stairs are constructed by old bricks with the era symbolic patterns, like walking in a historical space-time of the park. The people would have a strong historical apperception when walking towards the center and reach a climax of historical apperception by walking to the lawn center to touch the large-scale structures. This plot can be used for outdoor movie, music dry fountain, night light zone guiding indication to increase outdoor activities and dwell time of people in the park, serving to gather popularity in front of the museum and improve characteristics of the museum building.

中央草坪夜景
Night Scene of Central Lawn

时空大道　　鸟瞰图
Space-time Avenue　　Aerial View

设计理念：合理利用主干道，扩展主干道满足车行功能，因此将大门移至原门处，自南向北形成了旷阔的主轴，如一条时间轴将外来客引入园区内。大道的两侧有起伏的锈钢伴构筑的行走带，这也是时间轴的标示物品，上面镌刻了园区历史。行走带的高低表现1865的历史沉浮，从表面上看其是锯齿形的，与A1A2建筑相协调，而内在的含义是曲折而跌宕的，暗含了南京历史的文脉发展和地理基地建筑特征。

Design philosophy: utilizing the main street reasonably and extending the main street to satisfy the driving function; the front door is moved to the position where the original door is located to form an open and vast main shaft from south to north. It will introduce strangers to the park like an axis of time. At both sides of the street, there is fluctuant walking zone constructed by stainless, which is mark object of the axis of time and on which the history of the park is carved. The high and low walking zones express historical ups and downs in 1865. By seeing on the surface, it is zigzag shaped coordinating with A1 and A2 buildings, but the internal meaning is winding, waxed and waned, which impliedly means context development of history of Nanjing and geography as well as base building features.

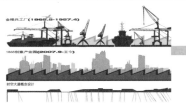

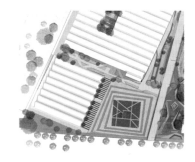

效果图
Design Sketch

学生感想
Students' Thoughts

经过三个月的时间我完成了此次毕业设计，毕业设计不仅是对前面所学知识的一种检验，而且也是对自己能力的一种提高。在此要感谢我的指导老师对我悉心的指导。在设计过程中，我通过查阅大量有关资料，与同学交流经验和自学，并向老师请教等方式，使自己学到了不少知识，也经历了不少艰辛，但收获同样巨大。在整个设计中我懂得了许多东西，也培养了我独立工作的能力，树立了对自己工作能力的信心，相信会对今后的学习工作生活有非常重要的影响。

I spent three months to finish the graduate project. It's not only the test of the previous knowledge we learned, but also the improvement of my own ability. Thank to my teachers for their careful guidance. In the design stage, I read a large amount of related material, communicated with experienced classmates, study by myself and asked questions to the teachers. It's a hard experience, but I learned a lot. During the design period, I've mastered lots of knowledge, improved my independent working ability, and established the confidence of my working ability. I believe it will further influence me a lot in my future study and working experience.

教师点评
Teachers' Comments

《穿越金陵2006》以时代变迁下的历史建筑与当下的人之间的关系为切入点，着眼未来，探讨建筑与人的共生。方案围绕A1、A2历史建筑所衍生的空间场，采用历史展示墙、下沉集聚草坪等手段力图创造连接历史与未来的景观空间，时空多元共生的情境，创意较好，景观塑造特色鲜明，但解决方案还可进一步完善。折线式元素的巧妙利用深化了场所记忆，并形成了统一的视觉美感。方案侧重形式，可增加针对园区工作者及游客的互动交流空间，每个节点上空间处理的合理性还需继续推敲。

——刘晓军 何方瑶

Across Jin Ling 2006, regarding the relationship between historic buildings and current people under the changing time as the point cut, discusses the co-existence of buildings and people with eyes on the future. The scheme focused on the space field derived from A1, A2 historic buildings attempts to create the landscape space connecting the history and future and the situation of multi-existence of time and space by adopting the history exhibition wall and subsiding gathered lawn, representing good creation and vivid landscape shaping, but the solution can make further perfection. The wise application of broken-line deepens visitors' memory and form a unified visual beauty; therefore, the scheme focuses on the form. The interactive communication space can be added for the staff in the park and visitors while the rationality of space design on each node requires further modification.

——Liu Xiaojun He Fangyao

CIID "室内设计 6+1" 2015（第三届）校企联合毕业设计
CIID "Interior Design 6+1" 2015(Third Scssion)University and Enterprise Joint in Graduation Design

工业遗产保护与展示设计
Industrial Heritage Protection and Display Design

C6
工业遗产保护与展示设计 Industrial Heritage Protection and Display Design

Industrial Heritage Protection and Display Design　工业遗产保护与展示设计

工业遗产保护与展示设计 Industrial Heritage Protection and Display Design

CIID "室内设计 6+1" 2015（第三届）校企联合毕业设计
CIID "Interior Design 6+1" 2015(Third Scssion)University and Enterprise Joint in Graduation Design

二等奖
Second Prize

文化蛹动机——1865 影工厂
Culture Worming Machine——1865 Film Factory

高　　校：北京建筑大学
College: Beijing University of Civil Engineering and Architecture
学　　生：刘凯南　郐嘉琦　姜帅
Students: Liu Kainan　Xi Jiaqi　Jiang Shuai
指导教师：杨琳　朱宁克
Instructors: Yang Lin　Zhu Ningke

关联页：126-131、190-198

刘凯南 Liu Kainan
郜嘉琦 Xi Jiaqi
姜帅 Jiang Shuai

逻辑架构
Logical Structure

A1 — 建筑改造 / 展示设计 → 理性实用 / 感性情绪

建筑层面改造成本巨大。不该浅显还原兵工厂或战争主题 该站在一定高度去审视。为长久留存成为经典.怎样审视？对于城市精神的概括;对于历史事件的态度;对于兵工文化的批判

The cost of remolding building level is huge. We should not restore the arsenal or war theme in plain, but gaze at it upon standing on a certain height. Make the history becomes classic. How to look at it? For the spirit of the general city; historical events in attitude; critique of culture industry.

通用展陈各分系统设计　各展馆特色互动设计　暗线内容展示设计

理性实用
Rational Pragmatism

人流分析
Analysis of Stream of People

主要人流有三，参观、工作、运货。依据人流进行需要入口的确定与后期设计工作。

Three kinds of stream of people mainly including: visiting, working and porter. To determine the entrance if necessary and the design work in later period according to the stream of people.

平面推敲
Plane Thinking

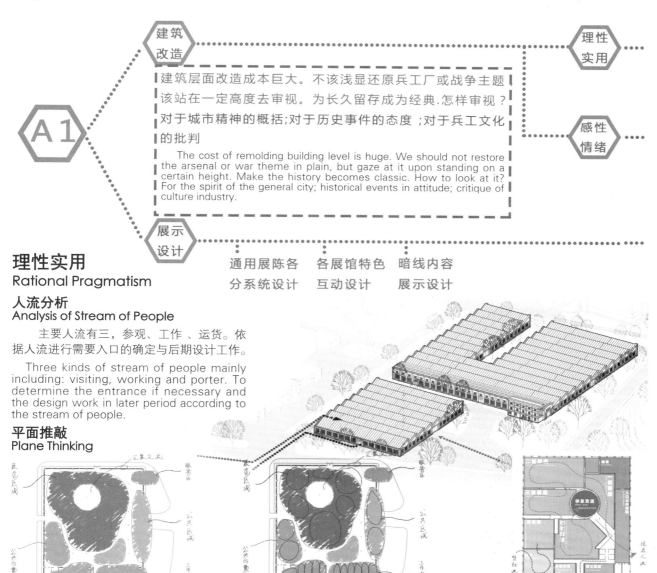

展览方式
Exhibition Way

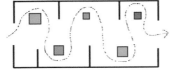

在展陈设计上我们选择串并联混合式展厅，在分展厅设计上选择单线浏览式。

On the design of exhibition and display, we choose the mixed exhibition halls in series and in parallel, and, on the design of branches of exhibition halls, we choose a single line browsing way.

感性情绪
Perceptual Mood

生生不息　坚韧不拔　重生的力量　动势　　　　　　　　　材质　　　　　　　　　空间体验

城市精神思路演化
City spirit idea evolution

哥特教堂簇拥向祭坛的动势让教徒感受到极大的心灵震撼，我的设计中延续这种具有贯穿感的动势，试图与城市精神有相似的情感冲动。

The kinetic potential of Gothic churches cluster round the sacrificial altar make Catholic felt great heart shock; we extend this kind of kinetic potential having the sense of passing through in the design to try to have the similar affective impulse to the city spirit.

"现在也不是凝固的" "不逃避老化，变旧。现在的历史应该被记录"用金属的氧化锈蚀记录这一过程。

"Today is not solidified", "We will not avoid aging and becoming old. The current history should be recorded", we record this process through corrosion by metal oxidation.

选择铁架为骨架，与实体的对比突出了新老部分不同，又隐喻着改建部分是形式噱头。内部是空的。内容的缺席需要老。

Choose iron stand as a skeleton to form a comparison with the entity, protruding difference between new and old parts and at the same time impliedly meaning that the reconstructed part is form stunt. The internal part is empty.

历史态度思路演化
Historical attitude idea evolution

老南京	今南京	明南京
伤痕残酷的过去	蓬勃发展的今天	以及膨胀的明天
The cruel past with scar of old Nanjing.	The flourishing development of the today's Nanjing	Tomorrow's Nanjing and inflated tomorrow

武器展厅歌颂人们的丰功伟绩，树碑立传，却忘记了武器带来的伤痛。展厅应该不止表现胜利的喜悦、复兴的希望；还有一面应表达深刻、悲伤、低调、批判的态度，空间体验应该是孤独绝望。引出蓝色参观暗线。

The weapon exhibition hall eulogizes great achievement of human, erecting a monument and writing biographies while forgetting pain brought by the weapon. The exhibition hall should not only exhibit the joy of winning and the hope of renaissance, but also hold the deep, sad, critical attitude in low profile. And the space experience should be lonely and despaired. The blue visiting dark line is thereby educed.

Industrial Heritage Protection and Display Design 工业遗产保护与展示设计

批判反思思路演化
Criticize and Rethink Idea Revolution

武汉乐队 AV 大久保，曾经准确地形容了他们对大都市的感觉。我把这段话引用过来说明我对南京的感受。
趋同的聒噪：城市一体化
客观角度：对于莫名的情感 只能陈述 不能做提出形而上的结果
嘲弄口吻：持反对意见，讽刺性
以上转化为空间体验，引入廊桥。

WuHan music Band AV Dajiubao once describes their feeling to the big city accurately. I introduce this passage to explain my feeling to Nanjing.
Convergent noisy: Urban integration
Objective perspective: can only state unexplainable emotion and cannot put forward super organic result
Mocking tone: holding opposite opinion, being ironic.
The aforesaid is transformed to space experience and introduced to Gallery Bridge.

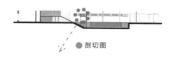
● 剖切图

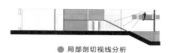
● 局部剖切视线分析

人走在其上经过暗线时的恐惧，经过铁架时消磨殆尽的耐性，留下焦躁与莫名其妙对大都市问题在情感上的客观还原。
People walk thereon, enduring fear when passing through the dark line and wearing down the depleted patience when passing through the iron stand, leaving impatience and confusing. This is the objective reduction of the problem of big city.

展示设计
Display Design

主题概述
Theme Overview

概念演化
Concept Evolution

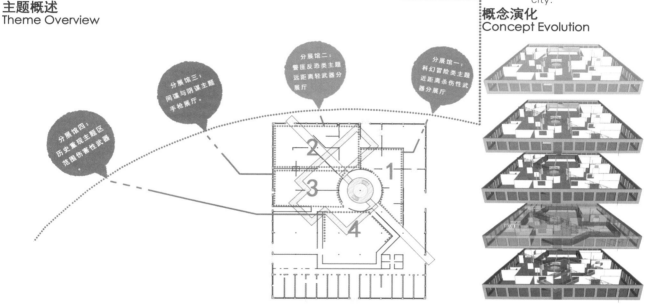

通用展示系统
Concept Evolution

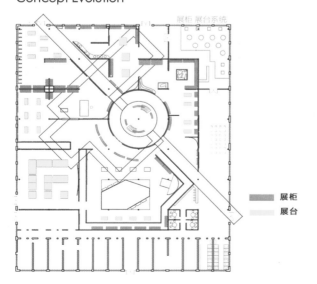

展柜
展台

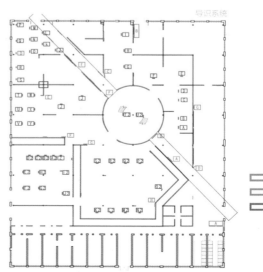

二级导识系统
三级导识系统
四级导识系统

193

导识系统
Guiding System

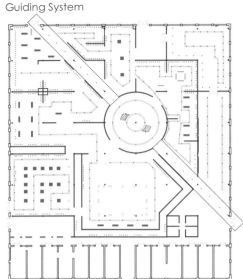

普通照明：

MR-16暗筒灯

局部照明：

吸顶式射灯

MR-16 格栅射灯，20w 或 50w，配光 24°，石英卤素光源

297*1197 无眩光高效格栅灯

597*597 无眩光高效格栅灯

特殊照明：

舞台效果灯

天幕灯

小型圆点射灯

General lighting:

MR-16 dark tube light

Local lighting

Ceiling type spotlight

MR-16grid spotlight, 20w or 50w, light distribution 24°, quartz halogen light source

297*1197 no glare efficient grille light

597*597no glare efficient grille light

Special lighting:

Stage effect light

Strip-light lamp

Small size dot spotlight

各分展厅设计
Design of the Respective Branched Exhibition Hall

科幻冒险厅展示脚本
The Science Fiction Adventure Hall Exhibition Script

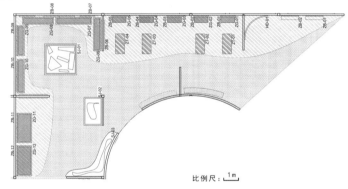

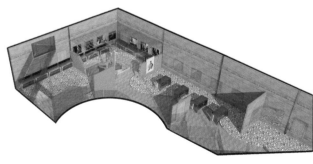

（1）．展台地铺：高反光金属材料，流动纹饰，与电影中场景契合。

（2）．墙面：带有模板印记的混凝土板。风格粗犷。

（3）．其他：1:1模型互动区，断壁残垣。碎裂的砖墙。

（1）.Exhibition stand paving: high reflective metal material, flowing ornamentation, according with scene in the film.

（2）.Wall: concrete slab with template stamp, rugged style

（3）.Other: 1:1 model interacting zone, dilapidated walls, cracked brick walls

警匪反恐展示厅脚本
Police Anti-terrorism Exhibition Hall Script

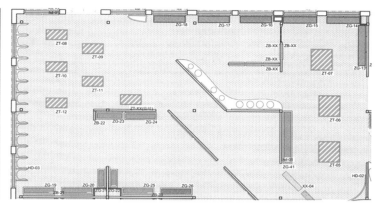

展厅鸟瞰图
Exihibition Hall Aerial View

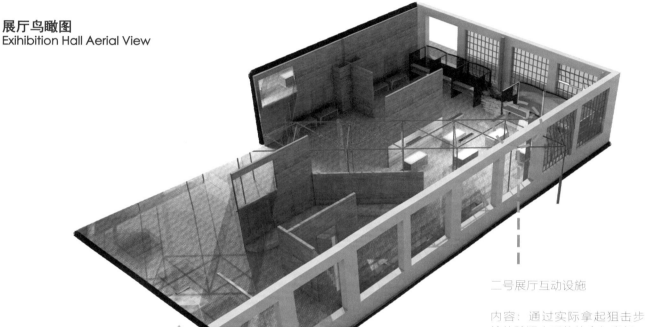

二号展厅互动设施

内容：通过实际拿起狙击步枪体验狙击手的使命与责任。

Two hall interactive facilities
Content: through the actual pick up the sniper rifle experience sniper's mission and responsibility.

形式：光电屏幕与真枪改造的狙击步枪模仿声音画面。得分计入每个游客的进入时拿的积分卡。

Form: photoelectric screen and genuine transformation of the sniper rifle to imitate the sound picture. Points included in the entry of each tourist to take the card.

间谍阴谋展示厅脚本
Spy Plot to Show Room Script

具体细分区域 展陈系统	区域 A	区域 B
一、电子显示屏	从第一部《007》到最后一部《007》精彩片段展示	《谍影重重》片段
二、二维展板	各部剧情的概要以及片名和主演名字。 位置:ZB-24	各部《007》中的手枪介绍,以及《谍影重重》中的手枪介绍图鉴中部的组装拼接台 位置:ZB-25 至 ZB-29
三、展柜	其最经典的 walther pp/ppk 手枪、walther p38/p1 手枪、walther p99 手枪 在 p99 旁边贴好 tips 上书《007》第 18 部明日帝国后半段 该款手枪取代了之前的老枪。 位置:ZG-27 至 ZG-29	walther p5、hk usp 手枪、sig p225 手枪、sig pro、陶喏思 pt945 手枪 细节拆解。 位置:ZG-20 至 ZG-32
四、展台	walther pp/ppk 手枪、walther p38/p1 手枪、walther p99 手枪拆解图 位置:ZT-13	《007》中 walther pp/ppk 手枪、walther p38/p1 手枪、walther p99 手枪、以及《谍影重重》中的 walther p5、hk usp 手枪、sig p225 手枪、sig pro、陶喏思 pt945 手枪的实物手枪拼装尝试,可定期举行观众参与的比赛,赢着可以把积分存入游客卡中。 位置:ZT-14 至 ZT-19
五、多媒体互动	附有类似杜莎夫人蜡像馆的代号 007 的主演以及名车阿斯顿马丁。可以合影留念。 位置:HD-03	
六、电影内容	《007》全系列	《007》全系列 《谍影重重》
七、枪械	walther pp/ppk 手枪、walther p38/p1 手枪、walther p99 手枪	walther p5、hk usp 手枪、sig p225 手枪、sig pro、陶喏思 pt945 手枪

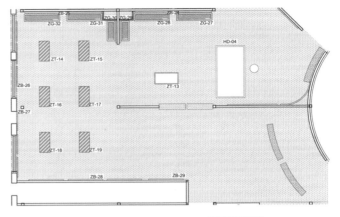

三号展厅互动装置

内容:现场主持与观众互动答题,胜者有机会与邦德的座驾一起合影。

Three Hall interactive device Content: the site hosted with the audience interactive questions, the winner has the opportunity and Bond's car with a photo.

形式:现场放置阿斯顿马丁复制道具车,满载高科技武器。金属感,科技感爆棚。

Form: the scene placed Martin Aston copy props, full of high-tech weapons. Metal feeling, sense of technology bursting.

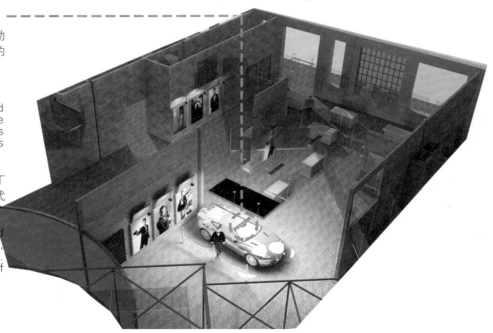

历史重现展示厅脚本
History Reproduction Exhibition Hall Script

具体细分区域 展陈系统	区域 A	区域 B
一、电子显示屏	《拯救大兵瑞恩》电影片段	《黑鹰坠落》
二、二维展板	剧情摘要，以及主要枪支的技术规格介绍，包括其中的小知识介绍，以及tips等相关小的知识点的介绍。 位置：ZB-34 至 ZB-38	剧情摘要，以及主要机械的技术规格介绍，包括其中一些小历史，小tips 等相关战争历史的介绍。 位置：ZB-39 至 ZB-43
三、展柜	"斯普林菲尔德" m1903 式步枪 m1 加兰德半自动步枪 勃朗宁 M1919 系列机枪 MG42 通用机枪（二战名枪）相关枪械的展示 位置：ZG-33 至 ZG-35	M249 轻机枪 以及 RPG 火箭筒实物展柜展示，以及解剖切面图等。 位置：ZG-36 至 ZG-40
四、展台	"斯普林菲尔德" m1903 式步枪 m1 加兰德半自动步枪 勃朗宁 M1919 系列机枪 MG42 通用机枪（二战名枪）实物，可触摸。 位置：ZT-29 至 ZT-30	黑鹰直升机结合人形模型加之场景复原，极力模仿电影中，历史中黑鹰直升机坠落时的场景。四周散落着各种电影系武器，包围中部的大型直升机场景，组合形成复杂的场景模式。 位置：ZT-31 至 ZT-32
五、多媒体互动	重型机枪体验装置，可以攻打虚拟目标然后完成一轮后 摄像头自动摄取其中精彩画面，在最后离开时，会转给参观者。 位置：HD-07	黑鹰坠落著名场景，黑鹰坠落瞬间，场景复原，人们可以通过中间的沟壑中走过，触发声音烟雾，全方位的模拟电影中的体验 而旁边的触摸式显示屏可任通过手指操作选择展馆中的介绍的电影观看。 位置：HD-06 及 hd-04 至 07
六、电影内容	《拯救大兵瑞恩》	《黑鹰坠落》
七、枪械	"斯普林菲尔德" m1903 式步枪 m1 加兰德半自动步枪、勃朗宁 M1919 系列机枪、MG42 通用机枪	M249 轻机枪、RPG 火箭筒

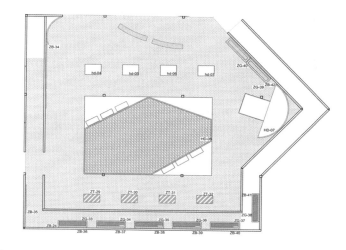

展厅鸟瞰图
Exhibition Hall Aerial View

三号展厅互动装置
Interactive devices of No. 3 exhibition hall

内容：走过黑鹰直升机坠落现场下的战壕，触发声光电烟雾等效果。
Content: walk through trenches on the spot where Black Hawk helicopter crashed, triggering sound, light and smoke effects.

形式：1:1 等比破损的黑鹰武装直升机，以及战壕中的真实的土地都加深模拟场景的真实感。
Form: 1:1 equal ratio damaged Black Hawk gunship helicopter and real soil in the trenches deepen the sense of reality of the simulated scene.

四号展厅互动体验二
Interactive experience II of No. 4 exhibition hall

四号展厅互动体验三
Interactive experience III of No. 4 exhibition hall

内容：人们通过操作互动体验台进行影片以及枪械知识的了解。
Content: people understand the film and firearms knowledge by operating the interactive experience stage.

形式：硬件为可触摸大屏幕，内嵌基于通用操作系统的软件运行环境，保证运行流畅，满足操作体验。
Form: hardware is a tangible big screen embed with a software operating environment based on a general operating system, ensuring smooth operation and satisfying operating experience.

暗线展示
Dark Line Exhibition

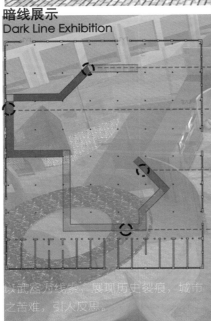

第四主题：南京大屠杀
1937年，淞沪会战与南京保卫战之后，日本侵略军的暴行。
No. 4 theme: Nanking Massacre
In 1937, savage act of Japan aggressor troops after battle of Shanghai and defense of Nanjing.

第三主题：太平天国屠杀
天国时，杨秀清与洪秀全内讧，与曾国藩湘军攻破天京后的屠杀。
No. 3 theme: the Taiping Heavenly Kingdom slaughter
At Heavenly Kingdom, Xiuqing Yang and Xiuquan Hong have internal conflict, the slaughter after breaking through Tianjing together with Hunan army of GuofanZeng.

第一主题：壬午之难
金川门之变，建文四年（1402）六月燕王朱棣直逼南京。明太宗靖难之役篡位后，屠杀建文帝幕僚及其亲属。
No. 1 theme: Ren Wu difficult
The door of the Jinchuan, four years (1402) June Jianwen Yan Zhu Di to Nanjing. Ming emperor after the battle of the usurper, killing the emperor staff and their relatives.

第二主题：明初四大案
洪武四大案，明初，朱元璋为整治吏治惩治贪污而策划的四大著名案件
No. 2 theme: four major cases at early Ming Dynasty
Hongwu four major cases, the four famous cases plotted by Yuanzhang Zhu at early Ming Dynasty for renovating local administration and punishing corruption.

暗线的室内设计紧扣与历史裂痕内容，以视频和二维展板的形式展现南京城市历史的裂痕。与外面明线展览对比强烈。

The interior design of dark line grips and tiles the historical crack to exhibit the crack of Nanjing history in the form of video and two-dimensional display boards, having a strong comparison with the external open line exhibition.

Industrial Heritage Protection and Display Design 工业遗产保护与展示设计

CIID"室内设计6+1"2015(第三届)校企联合毕业设计
CIID "Interior Design 6+1" 2015(Third Scssion)University and Enterprise Joint in Graduation Design

二等奖
Second Prize

机枪博物馆
Machine gun Museum

高　　校：	南京艺术学院
College：	Nanjing University of the Arts
学　　生：	许超　刘佳俊　杨广智　杨鸣
Students：	Xu Chao　Liu Jiajun　Yang Guangzhi　Yang Ming
指导教师：	朱飞
Instructors：	Zhu Fei

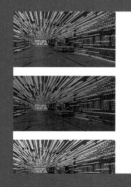

关联页：199-202

工业遗产保护与展示设计 Industrial Heritage Protection and Display Design

许超	刘佳俊	杨广智	杨鸣
Xu Chao	Liu Jiajun	Yang Guangzhi	Yang Ming

设计说明
Design Explanation

本展览馆不是单纯的从介绍各种机枪的角度出发，而是根据各种机枪的不同特性，探索运用特定的"空间场景感受"进行营造与诠释主题。尝试用意象表达主题空间气氛和深化主题，让观众在各个主题展馆体验和感受不同空间所营造出的"主题空间感受"。

希望借这种展览的方式表达主题内涵和吸引更多的观众来了解和体验、感受。以建筑空间元素与艺术作品语言进行有效的互动和延续，使参观者不仅与作品交流，也与该馆本身的空间艺术进行交流对话。

This exhibition hall does not start purely from an angle of introducing various machine guns, but explores and uses the specific "space scene feeling" to build and interpret the theme. We try to express the theme space atmosphere and deepen the theme by images, so that the audience can experience and feel the proper "theme space feeling" created by different spaces in each theme exhibition hall. By this kind of exhibition manner, we hope to express the theme connotation and attract more audience to know experience and feel. To make effective interaction and continuation with the building space element and artistic work language, as such, the visitors not only communicate with works but also make conversation with space arts of the hall per se.

一层路线图

二层路线图

Industrial Heritage Protection and Display Design 工业遗产保护与展示设计

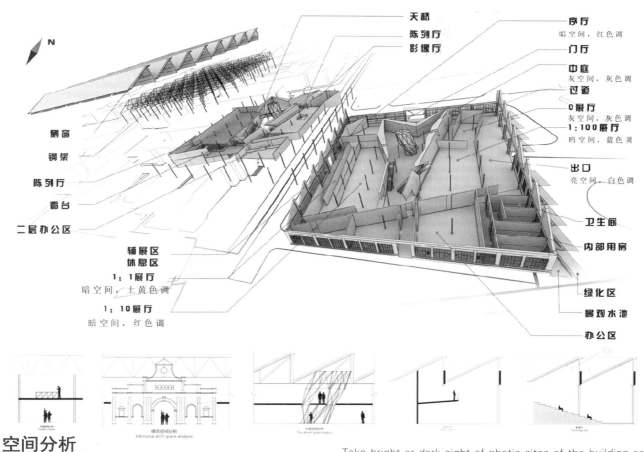

空间分析
Spatial Analysis

以建筑受光部位的强弱作为空间划分的重要依据，结合柱网，进行合理设计。把办公区与展区明显区分，把受光面较强的东面作为出口，让整个展厅上与下、明与暗、动与静有了较大的对比，不仅丰富了展示空间层次，也极大地提高了展示的趣味性和互动性。

Take bright or dark sight of photic sites of the building as important evidence of space dividing. Make reasonable design by combining with column grid. The office area is evidently differentiated from the exhibition area. The east, which has a strong photic face, serves as an exit to cause a very big comparison between up and down, brightness and darkness, and movement and quietude. This not only enriches the exhibition space level, but also greatly improves exhibition enjoyment and interactivity.

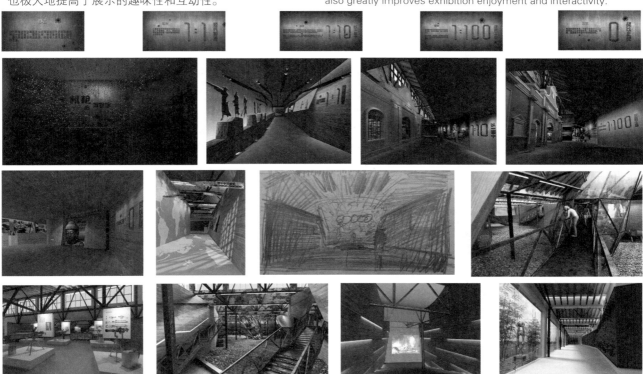

学生感想
Students' Thought

随着毕业日子的到来，毕业设计也接近了尾声。经过一年的奋战我们从南京到西安再到广州，有如下体会：在没有参加"6+1"以前觉得毕业设计只是对这几年来所学知识的单纯总结，但是通过这次活动发现自己的看法有点太片面。毕业设计不仅是对前面所学知识的一种检验，而且也是对自己能力的一种提高。通过这次互动使我明白了自己原来知识还比较欠缺。自己要学习的东西还太多，以前老是觉得自己什么东西都会，什么东西都懂，有点眼高手低。通过这次毕业设计，我才明白学习是一个长期积累的过程，在以后的工作、生活中都应该不断的学习，努力提高自己知识和综合素质。

在这次校企联合毕业设计中也使我们的同学关系更进一步了，与其它学校同学之间互相帮助，有什么不懂的大家在一起商量，听听不同的看法对我们更好的理解知识，所以在这里非常感谢帮助。

总之，不管学会的还是学不会的确觉得困难比较多，真是万事开头难，不知道如何入手。最后终于做完了有种如释重负的感觉。此外，还得出一个结论：知识必须通过应用才能实现其价值！有些东西以为学会了，但真正到用的时候才发现是两回事，所以我认为只有到真正会用的时候才是真的学会了。

As the graduate date draws near, the graduate project is coming to the end. According to the one year's work, we traveled from Nanjing to Xian, and then came to Guangzhou. Before taking part in the "6+1" joint project, we thought that the graduate project was only the summary of the knowledge we had learned. However, after joining the activity, I found that my previous opinion was too one-sided. The graduate project is not only the test of the previous knowledge, but also the improvement of my own ability. The interaction makes me realize that I am still short of knowledge. There are many things that I need to learn. In the past, I felt that I understood everything, and thought myself above my business. According to the graduate project, I began to realize that the study is the process of long-term accumulation. I should keep studying in my future work and life to improve my knowledge structure and comprehensive quality.

The College & Enterprise joint graduate project makes the relations with my classmates get further. We helped with each other. If there was something difficult to understand, we discussed with each other. Listening to the different opinions could make us understand the knowledge better. So here I express my appreciation for their help.

Anyway, whether I mastered the knowledge or not, there existed many difficulties in reality. It's difficult in the beginning. I didn't know how to start. After I finished it, I felt relaxed. Besides, there is conclusion: only the application could make the knowledge valuable. There is something that you think you've mastered. However, when you try to use it, you find you haven't mastered it. In my opinion, you really master the knowledge only if you can use it in practice.

教师点评
Teachers' Comments

南京艺术学院艺术与设计专业许超等4位应届毕业生，围绕南京晨光1865创意产业园A1馆环境设计，以"枪林弹雨"为主题，通过调研踏勘、创意策划，侧重在工业遗产保护与展示设计专题，完成了"机枪博物馆"作品。方案过程所做场地分析、内容策划、文案分析等较充分，在空间改造中，进行了空间设计、展线设计、自然光分析、氛围分析、导识设计、形象设计等。作品主题明确，反映出具有较好的专业基础功和专业实践应用能力，设计协同较好。

——陈静勇

Xu Chao and other 3 graduating students of Nan Jing University of the Arts, major in arts and design, focusing on the environment design of A1 pavilion in Nanjing Chenguang 1865 Creative and Technology Park, with the theme of "Heavy Gunfire" under the project "protection and display design of industrial heritage", have completed the work "machine gun museum" through research and on-spot investigation as well as creation and planning. The site analysis, content planning and plan analysis descried in the scheme are abundant. And in the space renovation, there are space design, exhibition line design, natural light analysis, environment analysis, guiding mark design and image designed. The theme of the work is definite, which reflected their solid professional basic skills and practical ability as well as good cooperative design.

——Chen Jingyong

专家点评
Experts' Comments

"机枪博物馆空间改造"，创作者处理这个题目的方式是直入主题，以"机枪"为题展开一场浓烈的想象，在这条线索上结合了现有空间的特质去编故事、做场景，融合了当下诸多时髦的展陈方式，如声、光、电的借用，如互动参与的考虑，在空间和氛围营造上展示了一定的素质。

设计师处处"精心"的设计制造了魅力也局限了设计，满满当当的内容未能为当下人的使用、新的活动内容或业态的进入留有余地，未能给未来的发展留有余地，"未尽"的设计和"留白"的处理会为设计带来更多可能性。

——庞伟

"The Reconstruction of The Machine Gun Museum Space", the author's way to deal with the subject is to go directly to the theme. He used the machine gun as the theme to spread a strong imagination. According to this clue, he made stories and scenes combined with features of current space. What's more, he integrated some fashionable exhibition methods, such as: sound, light, and electricity borrowing, interactive participation. It shows his quality in the space and atmosphere creation in a certain degree.

The designer's elaborate design creates the charm, but it's limited. There is not enough room for the people's use, new activities or business activities because of the full content. In other words, there is not enough room for the future development. The unexhausted design and blank process will bring more possibilities for the project.

——Pang Wei

Industrial Heritage Protection and Display Design　工业遗产保护与展示设计

CIID "室内设计 6+1" 2015（第三届）校企联合毕业设计
CIID "Interior Design 6+1" 2015(Third Scssion)University and Enterprise Joint in Graduation Design

三等奖
Third Prize

渠水流方
Introducing the Flowing Water to Interior Design

高　　校：	西安建筑科技大学
College:	Xian University of Architecture and Technology
学　　生：	刘璧凝　孙宗藜　李肖路　潘岩
Students:	Lliu Bining　Sun Zongli　Li Xiaolu　Pan Yan
指导教师：	刘晓军　何芳瑶
Instructors:	Liu Xiaojun　He Fangyao

关联页：106-111、203-207

| 刘璧凝 | 孙宗藜 | 李肖路 | 潘岩 |
| Liu Bining | Sun Zongli | Li Xiaolu | Pan Yan |

A1 设计理念
A1 Design Concept

展示空间是一个三维体，但由于参观者在展览空间中的活动特点，使其具有"五维空间"特性。参观行为是一种过程性活动，这种过程性参观活动都有一个时间的起止，因此展览场馆空间又有第四维要素——时间。参观者在展览空间中是通过游走的过程来感知视觉效果的，随着参观位置的移动来观察内部不同部位的空间而产生的不同的感受，这种感知过程是一种心理体验，因此展览空间具有第五维要素——心理。在此基础上，我们提出第六维要素——精神。这与博物馆的历史文化及人在参观时的想法密不可分。

The exhibition space is a three-dimensional body, but in view of activity characteristics of tourists in the exhibition space, the exhibition space has the characteristic of "the five space". The visiting behavior is a procedural activity. This kind of procedural visiting activity has a start-stop of time. Thus, the exhibition hall space further has a four-dimensional element—time. The visitors perceive visual effect by walking in the exhibition space and observe the space at different internal parts accompanying with movement of visitors' position to generate different feelings. This kind of perception process is a kind of psychological experience. Thus, the exhibition space has a five-dimensional element—psychology. On the basis of the same, we put forward a six-dimensional element—spirit. It is inseparable with historical culture of the museum and people's thinking when visiting.

A1 空间生成
A1 Space Generation

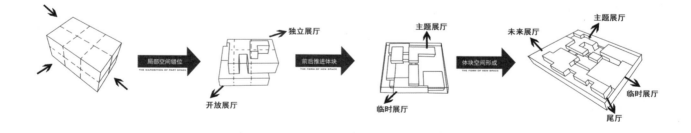

观影区 Film-viewing zone

序厅 Preface hall

卷轴展示区 Scroll exhibition zone

A1 一层 二层 功能流线分析
A1 Ground Floor, Second Floor, Function Streamline Analysis

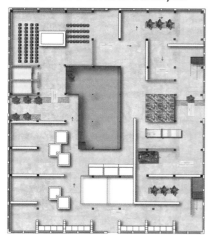

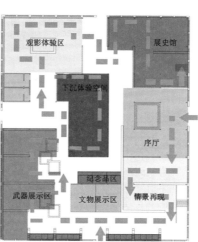

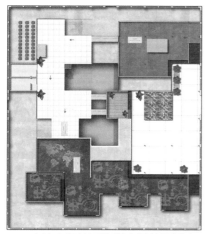

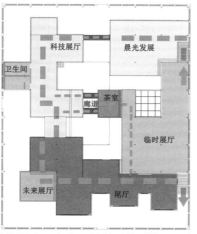

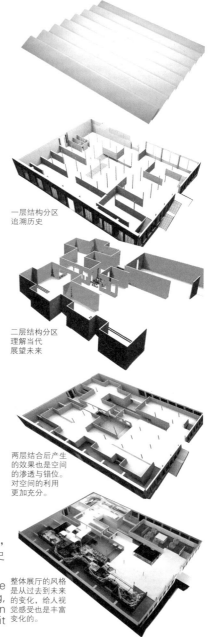

一层结构分区
追溯历史

二层结构分区
理解当代
展望未来

两层结合后产生的效果也是空间的渗透与错位。对空间的利用更加充分。

整体展厅的风格是从过去到未来的变化，给人视觉感受也是丰富变化的。

　　A1建筑内部的展览顺序是按照洋务运动的军事历史发展的前后排列的，经过设计，现在变得更加完美。而新的展区设计表达了对历史的尊重保护之场所精神，体现历史文脉。

　　A1The order of exhibition in the building is arranged in accordance with before and after military history development of Westernization Movement. By designing, it becomes more perfect. The new exhibition uses easily understood and even reversible processing method to express respect to the history and preserve spirit of the place to embody historical context.

展示手段示意图
Exhibition Schematic Diagram

多媒体技术必须服务和服从于展览主题及内容传播的需要，它应以内容解读为基础。如果过多地把科技设备作为重点，以技术玩花架子，脱离内容的技术应用反而适得其反。真正的灵魂是整个多媒体交互展示设计的创意思维和表现方式而不是一堆机器。博物馆与国际接轨，首先应该是创意地展示设计理念，而非单纯地引进科技设备。

The multimedia technology must serve and obey to need of exhibition theme and content propagating. It should take content interpretation as basis. If the technological apparatus is excessively taken as the key point to make a false and nice appearance by means of technology, the application of technology separated from contents is counterproductive. The real soul is creative thinking and manifestation mode of the whole multimedia interaction exhibition design instead of a pile of machines. The museum in line with international standards means that the creative exhibition design concept should be in line with the international standards rather than simply introducing science and technology equipment.

展厅内部的顶、地、墙界面；基本陈列设备如展柜、展墙、展架、展台、支架和道具等；辅助展品如文字图片、（电子）地图、图表、模型、沙盘、景箱、场景、蜡像、灯箱、声像资料编辑和制作；美术作品如壁画、历史画全景画、雕塑等；高科技装置如多媒体、动画、4D影院、环幕电影、幻影成像、观众参与装置等；特殊照明设备；文物征集和修复；等等。这一部分是从一层到二层的过渡部分，展现从过去到现代的过渡时期。展示手段为翻页式的屏幕航天产品展示空间，采用5D技术，让人能够亲自体验一下科技的发展。从第三、四、五、六个维度来与人与国际接轨。航天飞机零件，从客机到战斗机等。尾厅采用高科技的展示手段，主要展示待研发的新产品和创新的军事武器等。

Basic decoration of ceiling, ground and wall space in the exhibition hall; basic exhibition apparatus, for example: showcase, show wall, show frame, exhibition stand, support stand and stage property and etc.; auxiliary exhibits, for example: character, picture, (electronic) map, chart, model, sand table, view box, scene, wax statue, light box, audio-visual material editing and producing; artwork, for example: wall painting, historical panorama, sculpture and etc.; high-tech device, for example: multimedia, cartoon, 4D cinema, circular-screen movie, spectra vision, audience participation device and etc.; special lighting apparatus: cultural relic collecting and repairing and etc. This part is a transitional part from the ground layer to the second layer, and also a transitional period from past to now. Exhibition means: using aerospace products having a screen of page turning type to exhibit space, adopting 5D technology to make people experience the development of science and technology by themselves. From the third, fourth, fifth and six dimensions, the human is in line with international standards. The shuttle parts, passenger plane to fighter aircraft and etc. The end hall adopts high-tech exhibition means to mainly exhibit researched and developed new products and creative military weapons and the like.

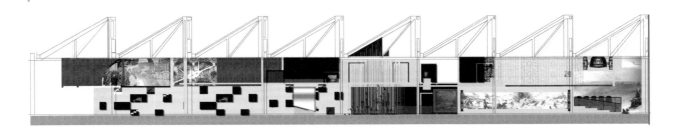

博物馆入口设计
Entrance Design of the Museum

博物馆的入口采用玻璃夹层水幕设计，流动的水体代表了流逝的历史，经过时间的冲刷，现在的晨光工业更加的"自强，求富"弥补了洋务运动的失败。新立面包住旧建筑，与原有的不透明旧建筑形成鲜明的对比。后者代表了旧社会下国人探索新道路的失败，是一种不成熟的道路。前者代表了现代社会的透明与先进，在当代人的努力下，现在的工业发展走向成熟，创造出属于自己的晨光梦、南京梦。入口空间附加在原厂房结构外侧，以不同的形体、材质加以区分，并通过连廊与相邻的厂房形成整体的空间形象。

The entrance of the museum adopts the design of water curtain sandwiched by glasses. The flowing water represents the elapsed history. Upon washing by time, the current Chenguang industry is more "self-improvement, fortune-seeking" to make up failure of Westernization Movement. New facade covers old buildings to form a striking contrast to the original opaque old buildings. The former represents transparency and progress of modern society. Under endeavor of the modern people, the current industry development becomes mature and creates own Chenguang dream and even Nanjing dream. The entrance space is attached to the outside of the original plant building structure and is differentiated by different shape and material, and forms the whole space image through the gallery and the adjacent plant building.

导示系统与武器展示
Guiding System and Weapon Exhibition

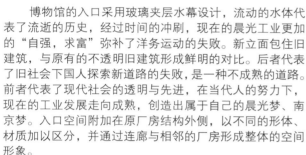

对金陵制造局的尊重与保护，在导识系统的设计上我们采用了一种工业形式的标牌，在锈蚀钢上镂空出文字形成一系列地面与墙面的导识系统，这种手法既运用了工业废料的再加工，也使整个材料与旧工厂的整个氛围相得益彰。简洁的导识系统带给人一目了然之感。原有材料的再利用，体现了对原有建筑的一种尊重和意境。

To respect and protect the Jinling Manufacturing Bureau, we adopt a sign in industrial form in the design of guiding system by hollowing out characters on the corroded steel to form a series of guiding systems of ground and wall space. This measure adopts reprocessing of industrial wastes and also makes the whole material and the whole atmosphere of the old factory complement each other. The brief guiding system will bring the sense of being clear at a glance to people. The recycling of the original materials embodies a kind of respect and conception to the original building.

工业遗产保护与展示设计 Industrial Heritage Protection and Display Design

CIID "室内设计 6+1" 2015（第三届）校企联合毕业设计
CIID "Interior Design 6+1" 2015(Third Scssion)University and Enterprise Joint in Graduation Design

三等奖
Third Prize

金陵风吟
Wind Sing of Jinling

高　　校：	西安建筑科技大学
College:	Xian University of Architecture and Technology
学　　生：	冯胤云　何其畅　韩杰林　黄超　胡铭中
Students:	Feng Yinyun　He Qichang　Han Jielin　Huang Chao　Hu Mingzhong
指导教师：	刘晓军　何芳瑶
Instructors:	Liu Xiaojun　He Fangyao

关联页：112-119、208-212

Industrial Heritage Protection and Display Design 工业遗产保护与展示设计

| 冯胤云 | 何其畅 | 韩杰林 | 黄超 |
| Feng Yinyun | He Qichang | Han Jielin | Huang Chao |

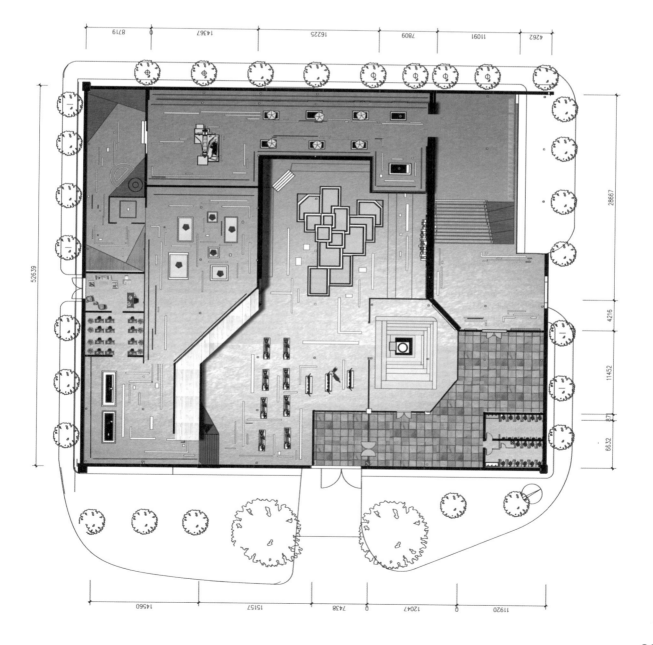

为了使参观者能更好的铺垫心境，在间流线设计上，我们采用了倒叙的叙述手法：先从一个建筑灰空间进入一个内置开敞的入口空间，在入口空间的墙上用浮雕展示金陵1865的发展史——宣告人们已进入展览馆；踏上步梯，展示晨光集团现有的一些科技产品，如其在航天航空事业上获得的各项国家颁发的奖项以及按比例缩小的各种航天航空器械模型，又如其在艺术造诣中精美的铜雕展品；下得两级台阶，进入一个突然昏暗的空间，里面没有任何展品——营造战争时期的灾难性毁灭与颓败，寓意战争时期任何人类的精神文明产物都将消失殆尽；随着地面、墙面数条灯带，将参观者引导进主展厅，其地面各有大凹面镜效果的展示窗口，可供人们俯视观赏被缩小的大型重机械兵器，甚至一些角度观赏是变形的、扭曲的，希望能够使人们明白战争会对人类世界造成扭曲与变形，并且主展厅可眺望下方的小型兵器展示区；通过不断向下的多媒体长廊，享受枪林弹雨的视听效果，了解兵器发展过程；进入小型兵器展示区，可以踩在小山似的展示台上，其里面的小型兵器也随之"踩"于脚下，正上方的钢架上垂挂着国画形式的战争图片，告诫人们和平是可以靠人类创建的，战争是可以被人类踩在脚下的；围绕着小型兵器展示区外，先是展窗与休息区相结合的展示体验区——复原了清朝、民国时期某个战争实况；然后是一面由电脑控制的可变形的展示墙；最后是多媒体历史教育厅，使参观者产生最强烈的战争反思。附属空间有设备的操作间及与参观内容配套的纪念品店。

For better mood of the tourists, we adopt a narrative skill of flashback on the streamline design: entering from a building gray space to a built-in open entrance space. On the wall of the entrance space, we present development history of Jinling 1865 with relief to tell people that they have entered the exhibition hall. By stepping on the stairs, some existing technological products of Chenguang Group are exhibited, for example, awards issued by the state in aviation and space industry and all kinds of aerospace equipments and models, that are proportionally scaled down, or elegant copper sculpture exhibits in artistic attainments. By stepping two down the stairs to enter into an abruptly dark space without including any exhibits, which means to build a catastrophic ruin and declining at the war time. It also means that any products of spiritual civilization of human will disappear. Walking along numerous strip lights on ground and wall space, the tourist is guided to enter into the main exhibition hall. By utilizing the exhibition window of each concave mirror effect on the ground, people can view the large-scale heavy machinery weapons by looking down. Even some angles for viewing are distorted and twisty to mean that the war will cause distortion and deformation; and, people can overlook the small-scale weapons exhibition zone in the major exhibition hall. Through the continuously downward multimedia gallery, the visitor can enjoy the seeing and hearing effect of rain of bullets and know the development process of weapons. Entering into the small-scale weapons exhibition zone, the visit can step on the exhibition stage which is like a hill and the small-scale weapon placed therein is also below the feet. The war picture in the form of traditional Chinese painting hanging down the above steel frame tells people that the peace can be established by people and the war can be grinded beneath feet by people. Around the small-scale weapons exhibition zone, there is first an exhibition experience zone where the exhibition window is combined with the rest area. It recovers a certain war real situation at Qing dynasty and the Republic of China era. And there is second a changing exhibition wall; at last, there is a multimedia history education hall, at there, the visitors can generate the most strong war introspection. The affiliate space has the space for operating the apparatus and the matched souvenir shop.

采选时间段
Selecting Time Period

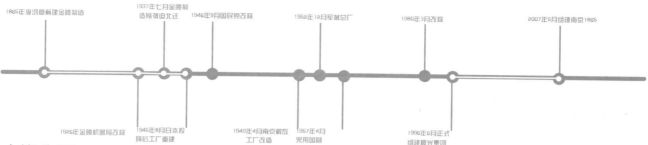

功能分区
Function Division

A1 展馆文案
Official Documents of A1 Exhibition Hall

A 序厅：历史背景及展馆展出内容（1865—1892）墙雕形式直观展示
B 展厅1：晨光集团发展史——（1996-2015）航天物品展示
C 展厅2：时光穿梭——（风格杂糅的空间）
D 主展厅3：金陵制造局（1865-1872）的兵器制造与发展过程
E 展厅4：（1872—1882）用视、听效果来演绎兵器发展过程
F 展厅5：战争还原（1882—1894）展窗与休息区相结合的展示体验区
G 展厅6：工厂机械、文物陈列（1872—1894）
H 展厅7：历史教育厅（战争的反思）
I 展厅8：临时展厅（作为艺术展览与艺术交流使用）纪念品店与设备间。

A Preface hall: historical background and exhibition contents of the exhibition (1865—1892) direct exhibition in the form of wall carving
B Exhibition hall 1: History of development of Chenguang Group——（1996-2015）aerospace objects exhibition
C Exhibition hall 2: time flies –(the space with blended styles)
D Main exhibition hall 3: Weapons manufacturing and development process of the Jinling Manufacturing Bureau (1865-1872)
4: (1872—1882) to interpret the development process of the weapons using seeing and hearing effect
F Exhibition hall 5: war reduction (1882-1894) exhibition experience zone where the exhibition window is combined with the rest area
G Exhibition hall 6: factory machinery and cultural relic arrangement (1872-1894)
H Exhibition hall 7: history education hall (rethinking of war)
I Exhibition hall 8: temperature exhibition hall (for use as art exhibition and art communication) souvenir shop and equipment room

 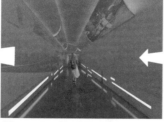  

工业遗产保护与展示设计 Industrial Heritage Protection and Display Design

A1 为兵器展览馆，展示品定位为金陵 1865 的前身金陵制造局时期国内外铸造的兵器和伴随历史印迹发展至今的晨光集团制造的一系列科技展品。展馆以时间流线的轨迹来设计展览空间流线，主体采用连廊走道式的空间串联形式，色调采用南京古建的黑白灰为基色，体现一个庄严肃穆的兵器类展览馆，并提取南京古建丰富的窗花元素融合厂房钢结构梁架材质设计展览馆的特色展墙、展柜。

A1 is exhibition hall of weapons, where the exhibits are positioned as weapons are cast at home and abroad at the time of Jinling Manufacturing Bureau—the predecessor of Jinling 1865 and a series of technological exhibits manufactured by Chenguang group developed so far accompanying with historical stamp. The exhibition space streamline of the exhibition hall is designed according to track of streamline of time, where the main body adopts the corridor and footpath in serial connection and the tone adopts black, white and gray of ancient buildings of Nanjing as basic colors to embody solemn weapons exhibition hall. The abundant window grille elements of ancient buildings of Nanjing are extracted to be combined with steel structure beam frame materials of the factory buildings to design the special show wall and showcase of the exhibition hall.

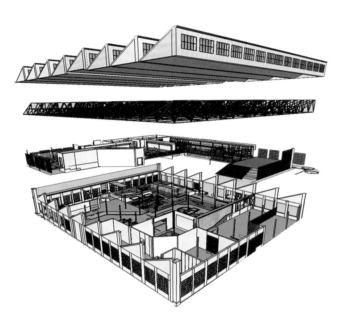

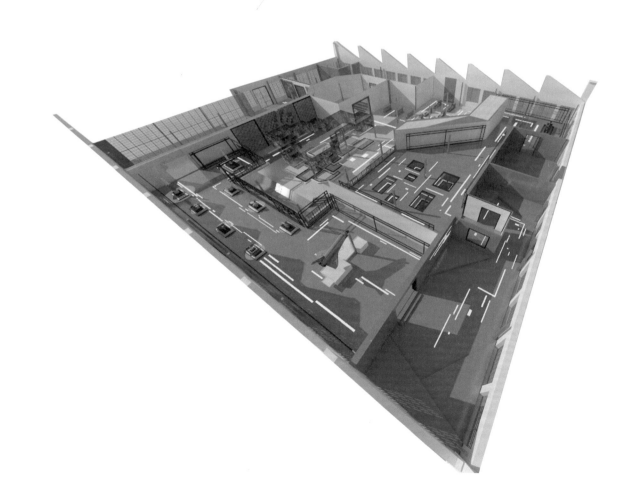

Industrial Heritage Protection and Display Design 工业遗产保护与展示设计

CIID "室内设计 6+1" 2015（第三届）校企联合毕业设计
CIID "Interior Design 6+1" 2015(Third Scssion)University and Enterprise Joint in Graduation Design

三等奖
Third Prize

城墙上的老机器
城墙下的老手艺
Old Machine On the Wall, Old Craft Under the Wall

高　　校： 北京建筑大学
College： Beijing University Of Civil Engineering and Architecture
学　　生： 顾文博　付晨辉　王曌伟　余祺盈
Students： Gu Wenbo　Fu Chenhui　Wang Zhaowei　Yu Qiying
指导教师： 杨琳　朱宁克
Instructors: Yang Lin　Zhu Ningke

关联页：120-125、213-215

工业遗产保护与展示设计 Industrial Heritage Protection and Display Design

顾文博	付晨辉	王墅伟	余祺盈
Gu Wenbo	Fu Chenhui	Wang Zhaowei	Yu Qiying

以兵工和时间为轴线，既展示以往经典，又像在倾诉故事。整个展馆贯穿情感设计，从迷惘到深沉再到无序而后穿过最混乱、畸形的时期焕然新生。这也契合着我们要把这种古城的城市精神带入其中的设计理念。

Taking ordnance industry and time as an axis shows the past classic and also seems telling stories, An emotional design runs through the whole exhibition hall, from confusing to deep and then to disorder and subsequently passing through the most chaotic and abnormal period to have a new life. This also conforms to our concept of bringing city spirit of this kind of ancient city to the design.

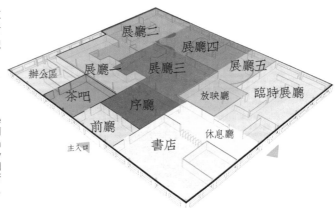

空间名称	面积	比重	可容纳人数	受众人群	采光要求	通风性	开放性	备注
前厅	100㎡	3.1%	40	游客	一般	强	强	
序厅	161㎡	5.0%	50	游客	一般	中	中	
历史厅	248㎡	7.8%	60	游客	良好	中	中	
晚清厅	414㎡	13.0%	120	游客	良好	中	中	
民国厅	348㎡	10.9%	90	游客	良好	中	中	
新中国厅	325㎡	10.2%	90	游客	良好	中	中	
未来武器	58㎡	1.8%	10	游客	一般	中	中	
环形影厅	72㎡	2.3%	15	游客	一般	中	中	
放映厅	131㎡	4.1%	40	游客	一般	中	中	
临时展厅	384㎡	12.0%	120	游客	良好	中	中	可租赁
休息厅	184㎡	5.8%	60	游客	良好	强	强	
厕所	56㎡	1.8%	15	游客、工作人员	良好	强	强	
书店	243㎡	7.6%	40	游客	良好	强	强	可租赁
茶座	146㎡	4.6%	35	游客	良好	强	强	可租赁
配电室	27㎡	0.8%	5	工作人员	一般	差	弱	
过道	85㎡	2.7%	15	工作人员	良好	中	中	
办公室	111㎡	3.5%	35	工作人员	良好	强	强	
总共	3193㎡	100%	840					24.2%可租赁

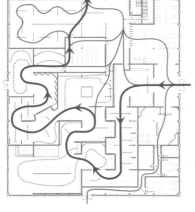

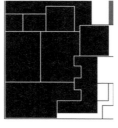

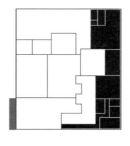

→ 主要动线
— 辅助动线
…… 游览动线

A1 的空间粗略地可分为主空间和辅空间，其比例约为 2∶1。

A1 的主入口设在东门，即靠近园区大门的方向，这样具有一定的仪式感。游客在游览 A1 时经过序厅，分别游览常设展厅、辅助展厅，到达临时展厅，而从北门出。

The main entrance of A1 is set at the east door, at a position close to the front door of the park. Such setting has a certain sense of ritual. When visiting A1, the tourist will pass through the preface hall, visit the standing exhibition hall and the auxiliary exhibition hall, and then arrive in the temporary exhibition hall to get out from the north entrance.

Industrial Heritage Protection and Display Design 工业遗产保护与展示设计

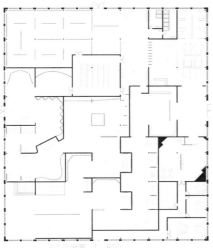

历史厅、晚清厅、民国厅与新中国厅介绍了晨光自1865年建厂以来的一系列重大历史事件和生产的各类重大轻重武器等。历史厅的尽头有一个穿越墙壁与天花的裂痕。这种裂痕造型，既可将游客引导至下一个游览空间，也寓意着动荡年代的开始。

The historical hall, late Qing dynasty hall, the republic of China hall and new China hall introduce a series of significant historical events and the variousgreat light and heavy weapons produced by Chenguang from establishment of the plant on 1865. At the end of the historical hall, there is a crack passing through the wall and the ceiling. This kind of crack modeling serves to introduce tourists to the next touring space and means a starting of the age of turbulence.

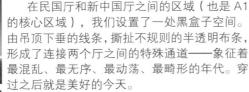
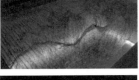
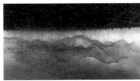

在民国厅和新中国厅之间的区域（也是A1的核心区域），我们设置了一处黑盒子空间。由吊顶下垂的线条，撕扯不规则的半透明布条，形成了连接两个厅之间的特殊通道——象征着最混乱、最无序、最动荡、最畸形的年代。穿过之后就是美好的今天。

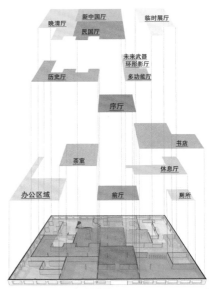

At the area between the republic of China hall and new China hall (which is also the core area of A1), we set up a black box space. Lines hanging down from the suspended ceiling tear irregular semitransparent cloth strip to form a special passageway connecting two halls, which represents the most chaotic, disorder, turbulent and abnormal time. By passing through, we meet the beautiful today.

工业遗产保护与展示设计　Industrial heritage protection and display design

CIID"室内设计6+1"2015(第三届)校企联合毕业设计
CIID "Interior Design 6+1" 2015(Third Scssion)University and Enterprise Joint in Graduation Design

三等奖
Third Prize

A2 综合服务区改造
Reconstruction of the Integrated Service Area

高　　校： 南京艺术学院
College : Nanjing University of the Arts
学　　生： 陆路　黄文进　陈飞宇　蔡文镇　刘曼羽
Students : Lu Lu　Huang Wenjin　Chen Feiyu　Cai Wenzhen　Liu Manyu
指导教师： 朱飞
Instructors: Zhu Fei

关联页：216-219

陆路 Lu Lu　　黄文进 Huang Wenjin　　陈飞宇 Chen Feiyu　　蔡文镇 Cai Wenzhen　　刘曼羽 Liu Manyu

创意阐述
Creativity Explanation

1865将业态多样化与新型消费方式相结合，结合建筑自身特点，保护性开发利用老建筑融入文化艺术、创意设计，在此基础上建设包含大厅、美术馆、主题展区、商铺等功能区域的综合服务区。

1865 factory combines diversification of format with new types of consumption patterns, the feature of the building itself and protective development and utilization of old buildings blended in culture art and creative design; on the basis of the same, the integrated service area including functional regions such as lobby, art gallery, theme exhibition area and stores are built.

设计理念
Designing Concept

在老建筑基础上，创造一个以展示为主的新型空间，强调新与旧的统一，追求科技、艺术与文化的结合。
（1）.扩大展示与体验空间的占比，强调商品的展示性：展示与体验性空间是整个改造最主要的空间，通过突出展示商品，体现商品的文化和艺术性，从而吸引游客，带动人气与消费，实现园区的品牌文化的宣传效果。
（2）.丰富空间层次，加强空间展示效果：①局部空间做高，丰富空间趣味性；②空间规划两条主轴线，两条辅线，强调轴线之间的关系；③通过材料的运用、色彩的搭配以及现代技术的融入，实现艺术与技术的统一。
（3）.科技在展示空间中的运用：①把最新科技融入老建筑中，注重空间的体验性；②互联网的运用，打破传统营销模式，实现"各家皆展馆"。

On the basis of old buildings, a new type space mainly for exhibition is created, which emphasizes new and old unification and pursuits the combination of technology, art and culture.
（1）.To Expand the Occupation of the Exhibition and Experience Space, and to Emphasize the Display of the Good: The exhibition and experience space is the main space of the whole reconstruction. It embodies culture and artistry of the good by highlighting the exhibited good to thereby attract customers and help lift sentiment and consumption so as to fulfill propaganda effect of brand culture of the park.
（2）.To Enrich the Space Hierarchy and Strengthen the Space Exhibition Effect: ① To Make the Local Space Higher and Enrich the Space Interestingness; ② Planning Two Main Axes and Two Auxiliary Lines in the Space and Emphasizing the Relationship between the Axes; ③ Unifying Art and Technology through Application of Materials, Matching of Colors and Blending of Modern Technology.
（3）.Application of Science and Technology in the Exhibition Space: ① The latest science and technology is blended into old buildings where the experience of space is focused on; ② Application of internet breaks the traditional marketing pattern to fulfill "every home is exhibition hall".

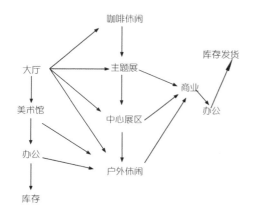

编号	主题	功能分区	平米	细分业态	平米	内容	创意亮点	功能特征
01		大厅	440			服务台、艺术展示、休息区		包含服务、接待、衔接美术馆、中庭、商铺、临展、茶吧等
02		主题展示	1100			展厅框架、展示性图文、液晶屏等	布局可变、灵动	用于临时性展览、发布会、宴会等活动的举办
03		美术馆	1200			展厅框架、活动轨道、展架、展板等	灰白简洁空间，运用滑轨、空间灵动多变	用于举办固定画展、工艺品展、艺术家专题展等
04	"陵"动——穿越1865	茶舍	400			触摸屏服务系统、工艺品展示、吧台座位	机械装置吊顶、艺术品展示与传统茶吧相结合	创意休闲、艺术展示
05		商业区	3400	商铺	2680	子空间、商品展示为主	"各家皆展馆"展示商品为主	以陶艺、布艺、书画、本艺等工艺品零售为主、中高档定位
				展示区	720	T台、展示框架、中心展示区	打破固有商业空间模式，引入T台与展架形式	T台与展示框架结合，中心区圆形框架结构灯光秀
				二层创意街区		文艺展示、创意小趣、创意服饰、工艺品等	与一层展示区形成看与被看的关系	空间布局灵活多变，以创意商品展示、销售为主、艺术气息浓厚
06		户外休闲	780			水池、树木、休闲座	围绕原有的植物做一个水池，形成户外聚集区	用于室内与室外间的联系，形成看与被看的关系
07		其他配套	1250	仓库	410			用于商品、展品的存储
				办公	260	子空间、用于创意办公	空间布局可变、灵动	主要用于个人工作室或小型企业办公
				卫生间	160			

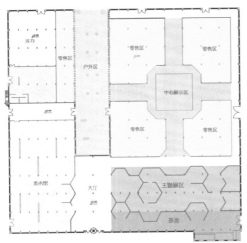

动与静（一阴一阳呼应）
Dynamic and Static (Cooperation of Yin and Yang)
蓝色为静空间
Blue is the sign of quiet space.
（美术馆、办公室、储藏与发货区）
(Art gallery, office, storage and goods delivery)
红色为动空间
Red is a dynamic space.
（中心展示空间、装置展示空间、商店、临时展示空间、聚会场所、唱吧、休息空间）
(Central exhibition space, equipment exhibition space, store, temporary exhibition, party place, sing bar and resting space)

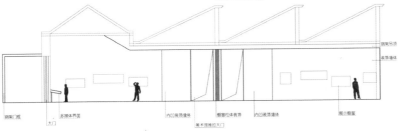
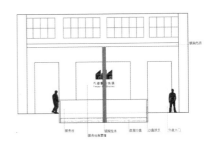

学生感想
Students' Thoughts

开始接到课题时却不知道如何下手。这次课题与大学四年中所学的知识略有偏差，在老师的指导鼓励下我们坚持一步一步往前走。曾艰难过，组员闹过，累过。随着时间的推移，我们毕业设计也接近了尾声。终于可以画上一个句号了！

很高兴我们有机会参与"室内设计6+1"校企联合毕设。知道我们和其他同学专业都是有差距的，但人们都说，设计是没有界限的。在做毕业设计过程中，经过现场调研、中期答辩等我们也发现了的目标的偏差。我们始终互相学习、交流。

经过了两个月的学习、交流，我们收获多多，也从中明显看到了自己的优点和不足，我们从中可以取长补短，不断地学习，不断地前进。

感谢我们的导师、活动主委、联合高校导师们，谢谢你们的一直陪伴。

When received the theme, I didn't know how to start the work. This project is not closely related to our knowledge we learned in the last 4 years in the college. According to our teacher's encouragement and guidance, we kept on going step by step. It's very hard. The members used to feel irritable and tired. As the time goes by, our project is coming to the end. We are glad that our works are finished.

We're glad that we could have chances to take part in the 6+1 Graduate Interior Design joint activity. We know that other students' level is higher than us. However, people say that the design is limitless. During the project period, we did research on site, replied in the midterm. After that, we found that our direction was deflected. We learned from each other and communicated with each other all the time.

According to the two-month study and communication, we benefited a lot, and found out our own advantages and shortages. We learned from each other, and kept on studying and moving forward all the time.

Thanks to our teacher, activity organizers, and teachers from joint colleges. Thank you for keeping us company all the time.

教师点评
Teachers' Comments

该组同学对1865整个园区环境调研深入、清晰，对A2馆的建筑空间能够从历史保护建筑角度出发，进行保护性开发利用室内空间，同时对室内通风与日照进行了科学有效的分析。

整个方案布局相对合理，"动区"与"静区"空间组织得当，交通流线清晰并符合人流疏散要求。空间装饰造型能够从整个园区的历史演变文脉提取出相应的元素符号，并加以抽象概括作为空间设计的装饰符号。整个空间的设计表现较为精致，材料选择与应用恰当，空间整体氛围具有一定的艺术表现力。室内空间装饰细部处理较为生动，室内外空间环境具有一定的有机联系，使得整个A2馆室内外环境设计与1865整个园区和谐统一。

由于A2馆建筑空间的特殊性，顶部造型较为复杂，该组同学对A2馆室内天花设计还缺少一定的详细组织设计，缺少更清晰的针对天光与人工光照可调节与控制天花设计措施。

——马辉

The group of students studied the whole 1865 Innovation Campus deeply, and the information is clear. In the view of the protective historical building, they made use of the A2 building's space to develop the inner space. Meanwhile, they scientifically analyze the ventilation and sunshine in the inner space.

The project is relatively reasonable. The "moving area" and the "silent area" are organized well, the traffic streamline is clear. It's proper to evacuate the crowds. During the space decoration, the students extract the symbol of element from the history development context in the whole campus, and abstract them as decoration symbol of the space design. The whole space is delicate. The material selection and its application are proper. The atmosphere of the space has some artistic expression. The process of the inner space decoration is vivid. There is some organic connection between inner space and outer space. That makes the whole A2 building inside & outside environment be in harmony with the whole 1865 Campus.

Because the space of the A2 building is special, the roof modeling is complex. The group of students is short of some detailed organization design. What's more, they are lack of the measure to adjust and control the ceiling aimed at daylight and illumination.

——Ma Hui

专家点评
Experts' Comments

整个方案思路清晰，结构完整，考虑到园区的历史文化底蕴，进而到空间布局和装饰，并紧紧与商业需求关联。工业元素和佛教元素以及建筑特色的提取非常有创意，具象化后的布局让整个A2变得可爱和灵动。

方案切合实际地提到了A2目前运营中存在的问题，并且注意到了将来的商业发展方向，如果在电子商务平台的部分，能够深入清晰地挖掘设计思路，那么将成为整个方案的点睛之处；而事实上，A2的实际运营过程中亦已经建立了相应的互联网交易平台。

——邹超

The design scheme is definite and has complete structure. In terms of the historical background of the park and further the space layout and decoration, the scheme meets the business demand closely. The industrial factors, Buddhist factors and the abstraction of architectural features are brilliant creative thinking. The embodied layout causes the A2 venue to be smart and lovely.

The scheme has mentioned the existing problems in the operation of A2 Venue in compliance with the fact and noticed the business development trend. If the author can unearth the design clues more deeply on the E-commerce platform part, it will be the shining point of the whole scheme while In fact; A2 venue has established the corresponding Internet trading platform.

——Zou Chao

CIID "室内设计 6+1" 2015（第三届）校企联合毕业设计
CIID "Interior Design 6+1" 2015(Third Scssion)University and Enterprise Joint in Graduation Design

联合毕业设计的教学探讨
Teaching of Graduation Design

Three relations and three principle
—— Teaching exploration of third sessions of "Interior design 6+1" University and Enterprise Joint in Graduation Design

The CIID "interior Design 6+1" University and Enterprise Joint Graduation Design have been held for three sessions since 2013, each year a design company assigns a topic. Seven universities participated in this activity this year. This activity aims at promoting the communication and the connection between enterprises and universities. The topics are different each year. In 2013, the topic of the first session was the after game design for the national stadium "Bird Nest"; for 2014, it was the environmental design of Shanghai subway rebuilding; and this year, the subject is Environmental design of Nanjing Chenguang 1865 Creative Industrial Park. As an instructor, I guided my students from Tongji University in participating in this event for three successive years, and received many thoughts and experience.

Three months of graduation design is rather short, yet it was an unforgettable experience for both the students and the teachers. Being fully devoted, we gathered our thoughts and confusions, and witnessed our growth and gains together. This event provided universities from all over China with a platform to present their confidence and talents, while groups of students were reporting their programs on stage, the jury teachers listened and made comments with full attention casting confirmative and encouraging looks towards the students. Facing this one and only teaching platform of interior design, of which the mechanism is maturing and the impact is increasing, we wonder if our teachers are ready to make changes from the aspects of teaching psychology, teaching methods and professionalism.
In my opinion, to successfully guide the graduation design, the instructors must coordinate well the following three relationships:

1. The relationship between variability and invariability.
Every year, the topics change, they involve different space environment types and requirements; the students vary too, some have higher abilities than others; in the meantime, the judges and award methods are different as well. For the guiding teachers, these changes that we mentioned above are dynamic factors. On the contrary, the design concepts that we formed over a long period of time, working attitude and guiding approaches are rather static factors, how to deal with the variables with invariables is a question that every instructor need to confront and think about.

2. The relationship between proposing and innovating a topic
The interpretation of a topic is an important part in the graduation design every year, all of the topics of the past three years have followed the principles which correspond to the hot issues of modern cities and the development of the society. From high-end city landmarks to daily circulation spaces of the general public; from single buildings to entire plants; from contemporary constructions to preserved architectures of the Republic of China, their type, volume, year of construction, key point in designing and requirements are all different. Thus requires the instructors to control clearly and redefine the content, range and significance of the topic in light of their own characteristics. Moreover, they need to fully excavate and collect powerful clues of field researches as first-hand information, so that they can have a profound understanding and reflection on the topics in the early period of a design, and then find the essential problem in order to provide basis and ideas for the next stage in the design.

3. The relationship between teaching and learning
The relationship between teaching and learning has been a basic but long-term topic in education; it is not an easy job to handle well this relationship. The best status is the teaching that benefits the students as well as the teachers, which requires the mutual efforts from them both. To do so, a good state of mind, a solid professional knowledge and a broad academic vision are the comprehensive qualities the teachers must possess. For students who are passionate about design, the guidance of a teacher only takes up 30%, the other 70% lies in the hardworking and the savvy of a student. But for those who are less talented and devote less, the role and devotion of the instructor appear to more important. plans will be largely enhanced.

三个关系和三个准则
——三届"室内设计 6+1"校企联合毕业设计的教学探索

左琰

Zuo Yan

CIID "室内设计 6+1" 校企联合毕业设计自 2013 年开始至今已举办三届了，每年一个设计企业命题，参与高校达到了 7 所，旨在促进企业与高校之间的交流和对接。三年选题各有不同：2013 年首届为国家体育馆"鸟巢"的赛后改造设计，2014 年第二届为上海地铁车站环境改造设计，2015 第三届年为南京晨光 1865 创意产业园区环境设计。我作为指导教师，连续三届带领同济学生参加，收获了许多感想和经验。

短短三个多月的毕业设计，对学生和指导教师来说都是人生中一段难忘的经历。全心的投入，一起思考和困惑，也一起见证成长和收获。室内设计 6+1 的平台给了全国各地高校展现自信和风采的舞台，当我们看到一组组青春洋溢的学生在台上汇报方案时，教师们则坐在台下的评委席上认真聆听和评点，并投以肯定和鼓励的目光。对于这个机制日趋成熟、影响力逐年增大的国内唯一的室内方向的毕业设计教学舞台，试问我们教师从教学心态、教学方法和专业能力等方面是否做好了相应的准备和改变？

在我看来，要指导好毕业设计，指导教师需协调好以下三个关系。

1. 关于变与不变的关系

选题每年都变，涉及不同的空间环境类型和要求，学生每年也在变，能力有高有低，评委和评奖方法每年也在变，那么对于指导教师来说，与上述动态的变化因素相比，我们以往长期积累形成的设计理念、工作态度和指导方式是一种较为稳定的静态因素，因此，如何应对这些动态因素，如何以不变应万变，是每个指导教师需要面对和思考的问题。

2. 关于命题的破与立的关系

命题解读是每年毕业设计指导的一个重要环节，三届的选题都以契合当代城市和社会的发展热点为原则，从高大上的城市地标建筑到普通民众的日常交通空间，从单体建筑到整个厂区，从当代建筑到民国保护建筑，建筑的类型、体量、建造年代等各不相同，设计重点和要求也不同，这需要指导教师根据自身特点来对命题的设计内容、范围和意义有清晰的把控和再定义，充分挖掘和收集基地调研的有力线索作为第一手资料，力求在设计前期对命题有全面深入的了解和思考，找出实质性问题，为下一阶段的解题提供依据和思路。

3. 关于教与学的关系

教与学是教育界一个基本而长期的话题，处理好这个关系并非容易。教学相长是一种最佳状态，这需要师生双方共同努力来完成。要做到这一点，良好的教学心态、扎实的专业能力、开阔的学术视野是教师需要具备的综合素质。对于设计有热情的优秀学生，教师的指导占三分，学生自我的勤奋加悟性占七分，而能力和投入都一般的学生，教师的作用和投入就显得尤为重要了。

The "Interior Design 6+1" graduation design for three successive years threw me a huge challege and a space to bring my advantage into full play. Facing the three relationships, I practiced the following three acting norms throughout the entire teaching process:

1. mental training stress on design research
Design thinking is a problem solving oriented way of thinking, it is applied to offer a practical and creative solution for a problem or an event that aims at improving results; trainings on design thinking which are led by researches are rational and rigorous; concepts that were generated under this thinking frame are supported by theories and data analyses, it refrains people from falling into extreme and personal ideas. In this way, a student's logical capability while making plans will be largely enhanced.

2. Forming an open and inclusive design vision
Even though classifications in majors such as urban planning, architecture, landscape design and environment are various, eventually their values and advantages need to be accomplished and reflected on by their integration in actual projects. Opening design vision and adjusting design psychology are guarantees for qualified designs; in fact many theories and methodologies are the same. Interior design major students need to understand a little about architecture, architecture major students need to know a few about urban design; urban planning major students need to get the picture of one or two things about historic preservation, environment major students need to be familiar with art and culture. Only when one is at a higher level can he has a broader view, sees further and stays calmer. "Interior design 6+1" joint graduation design organizes together universities with different professional backgrounds and disciplines from all over China to work on a design topic, this will certainly require the students to understand and think over the topics from multiple angles, and consequently lead to diverse results. Besides, three years of teaching practices have proved and enhanced the trend and value of having multiple answers for the same question.

3. Exploring bonding points with cultural and social focuses.
Undergraduate students tend to value the design over the theory, the form over the technique, the phenomenon over the essence. Therefore, it is important to reinforce the cultivation of humanistic ideology and social responsibility in graduation designs. Education in humanistic thinking is an indispensable part in architectural education; it is the spiritual source that infuses soul in a design work, accumulation of humanistic thinking not only improve a student's professionalism, but more importantly, it also provides the students with an important nutritive basis in forming correct values about their life and career. Furthermore, social responsibility is a basic quality students have to possess in modern time. In the early reading period of the graduation design, we encourage the students to open their mind, start from the actual needs in the society, concern about vulnerable groups, heighten their environmental awareness and embody humanistic care in the details of their design work. In this way, they can combine the topic of the graduation design with social focuses, which will also make their design thinking and their final result be rich in social and humanistic value.

As a bridge that connects universities and universities, universities and enterprises, the "Interior Design 6+1" School and Enterprise Joint Graduation Design have received more and more attention from people. I am very thankful for this platform set up by the academy and the universities; it allowed all the participants to confidently show their own charm and personality, and I am looking forward to another three years of wonderfulness.

连续三年的"室内设计6+1"毕业设计教学给了我很大的挑战和发挥空间，面对以上三个关系，在整个教学过程中实践着以下三个行动准则。

1. 注重设计研究的思维训练

设计思维是一种以解决问题为导向的思维形式，用于为寻求未来改进结果的问题或事件提供实用和富有创造性的解决方案。以研究为先导的设计思维训练是理性而严谨的，在这一思维框架下生成的概念想法有理论和数据分析作为支撑，避免概念陷入狭隘偏激、个人化的思想泥潭，这样可以使学生方案的逻辑能力大大增强。设计思维贯穿于毕业设计过程始终，包括对找出问题、分析问题及解决问题三个阶段的能力培养。正确的设计思维训练，可以使学生在毕业后走向工作岗位时掌握一套行之有效的解决实际工程问题的方法与途径。

2. 形成开阔包容的设计视野

尽管规划、建筑、景观、环境等专业学科分类繁多，但它们的价值和优势终需借助实际项目的整合得以完成和体现。开阔设计视野，调整设计心态，是完成优质设计的保证，事实上许多学科理论和方法体系是相同的。学室内的要懂点建筑，学建筑的要懂点城市设计，学规划的要懂点历史保护，学环境的要懂点人文艺术。只有站得高，视野开阔，才能望得远，从容淡定。"室内设计6+1"联合设计活动将全国各地不同专业背景和学科方向的高校组织在一起做一个设计课题，这势必会从多视角去理解和思考命题，最终也必将导出多样化的解题结果，三年的教学实践印证和强化了这一题多解的趋势和价值。

3. 探索与人文社会热点的结合点

本科生阶段学生往往重设计轻理论、重形式轻技术、重现象轻本质，因此，在毕业设计中加强人文思想和社会责任的培养非常必要。人文思想是建筑教育中不可缺少的部分，是为设计作品注入灵魂的精神源头，人文思想的积淀不仅使学生的专业素养得到提升，更重要的是为建立正确的人生价值观和职业观提供了重要的营养基础；而社会责任感是当代大学生应具备的基本素养，在毕业设计初期命题解读阶段，鼓励学生打开思路，引导学生在方案设计中从社会实际需求出发，关注社会弱势群体，增强环境保护意识，在细节中体现人性关怀，从而将毕业设计命题与社会热点紧密地结合起来，促使设计思考和最终成果富有较强的社会和人文价值。

"室内设计6+1"校企联合毕业设计作为高校与高校之间、高校与企业之间的桥梁正越来越受到人们的专注，感谢学会和各校合力构建了这个交流平台，让所有参与的师生带着满满的自信展现了各自的魅力和风采，期待下一个三年精彩继续！

Reflection on the "Interior Design 6+1" University and Enterprise Joint in Graduation Design

For a long time, the graduation design topics for students are selected by the instructors. However, as all kinds of exchange activities have become more frequent in recent years, the teaching method and mode for graduation designs have changed greatly. As one of them, the school and enterprise joint graduation design is expanding every day, and this teaching mode is becoming more and more popular among students.

The "Interior Design 6+1" School and Enterprise Joint Graduation Design hosted by CIID (China Institute of Interior Design) built a platform of high quality for students of this major, after three years of exchanges, the activity has received a good outcome. As one of the participating schools, students from School of Architecture of the South China University of Technology have gained a lot, and as their guiding teacher, I also have certain thoughts on the activity.

1. A topic of general interest, practice on a real project

Graduation design is a comprehensive assessment of the students' achievement in their undergraduate study; it is the last part of their courses in college. The selection of a topic is an essential part of the graduation design; it affects directly the students' passion, initiative and engagement in the activity.

Our school has participated in three sessions of "Interior Design 6+1" joint graduation design; we can see from the feedback that, generally, this activity is very appealing. However, during the on-campus application stage every year, for different topics, the reactions were different, good topics tended to be more popular and got more applications from students.

Usually, the topics we assign in class are fictitious; therefore, there are plenty of theoretic and abstract contents in students' work. These topics are often less difficult to design, as a result, the proportion of the students' subjective imagination in their design is rather high. For students, they won't be able to learn the method to solve real problems from fictitious topics. The analyses and deliberations appear to be logical, yet the final results are often like castles in the air, which are unable to be carried out, as there are too many flaws.

Each year, the topic of "Interior Design 6+1" school enterprise joint graduation design is selected by the enterprises in the society, these topics were often the hottest issues at the time, they were real projects, which can stimulate the students' initiative of participation, in the mean time, it allowed us to combine the theory with social conscious, add more practical value in the graduation design, and enhance the students' awareness of social hot affairs.

In addition, real projects can provide the students with actual research and analyses; they would be able to experience the process by themselves, and learn to get hold of the solutions to real problems in the projects.

2. Seeking common ground while reserving differences and opening to diversification

Seven participant universities come from different regions of China, each one of them have different professional backgrounds, including architectural design, interior design, environmental design, engineering design and exhibition design majors. All the universities have a mature teaching system and their unique advantage, each one has their own educational philosophy, value and characteristic. Therefore, they will have different operational mode and design approach facing the same topic, and the final result will have great difference. The diversified feature can be found in multiple aspects such as regions, professions, teachers and students.

This kind of diversification and difference is the no other than the characteristic of "Interior Design 6+1" school and enterprise joint graduation design.

3. Accumulation of experts and broadening minds.

Regular graduation designs rely mostly on the guidance of instructors; however these teachers have certain limitations for the students.

We invited different experts and scholars to participate in the teaching during the opening of topic, mid-term defense and final defense of this activity. The crew is huge and it reaches an extensive level, the backgrounds of these experts are various, they all have rich experience and their own personality. With their own unique opinions, these experts, who are very influential in the field, can offer the students with abundant information, comments on their plan, reasonable advises and philosophical inspirations, which is a rare opportunity for students to learn and to communicate. By means of this platform, we are able to invite these experts into our classes, listen to their lecture and receive their support; this is

"室内设计 6+1"校企联合毕业设计活动随想

姜文艺
陈建华

Jiang Wenyi
Chen Jianhua

长期以来，毕业设计课题大多以导师选择的课题进行课题设计创作或研究。但近些年随着各种交流活动的日趋频繁，毕业设计教学的手段、模式也有了很多的改变，校企联合毕业设计作为其中一种模式正在日益扩展，这种教学模式受到了学生们的青睐。

由中国建筑学会室内设计分会（CIID）主办的"室内设计 6+1"校企联合毕业设计在这一学科中搭建起了一个优质的平台，通过三年的教学交流，活动取得了很好的成果。华南理工大学建筑学院作为参与活动的院校之一，学生有很多收获，作为指导教师也有一些感想。

1. 热点课题，真题实做

毕业设计是对学生本科学习成果的一次综合考核，是学生在校学习的最后一个环节。而毕业设计选题是毕业设计的关键环节，直接影响到学生毕业设计的热情、积极性与参与度。

从华工大参加三届"室内设计 6+1"校企联合毕业设计反馈来看，就校内毕业设计选题报名环节总体而言，"室内设计 6+1"联合毕业设计有较大的号召力，但每届不同的题目反响差异也有所不同，好的课题学生响应度高，报名人数多。

平时的课程设计作业有些假题，在学生作业当中概念性、抽象的东西较多，假题往往也降低了设计的难度，导致学生主观臆造的设计成分比例偏高，假题对学生而言无法学习到真正面对实际问题的处理方法，分析推敲貌似很有逻辑，但最后的成果往往有如空中楼阁，无法落地，存在很多硬伤。

"室内设计 6+1"校企联合毕业设计每年的课题都是由社会企业选取社会当下最热门的问题进行选题命题，题目都是真实的项目课题，这样的课题能激发学生们参与的积极性，同时可以让教学的理念结合社会意识，使毕业设计更有实际价值，又能增强引导学生关注社会重点问题的意识。

此外，真题能为学生提供真实的调研分析，体验实操过程。同时能让学生学习掌握面对实际项目问题的解决方法。

2. 求同存异，开放多元

参加"室内设计 6+1"校企联合毕业设计的七所院校分别来自全国不同的地区，且又有各自不同的专业背景，分别包括了建筑设计专业、室内设计专业、环境艺术设计专业、工业设计专业及展陈设计专业等，每个院校都有各自成熟的教学体系与优势，每个学校又都有着各自不同的办学理念、价值取向、教学特色，对同样的课题会有着截然不同的操作模式与设计方法，因此，最后成果也有很大的差异性。在地域、专业、教师、学生等多方面体现出多元化的特点。这种多元化、多差异的特点也正是"6+1"校企联合毕业业设计的特色。

3. 专家汇集，广开思路

常规的毕业设计大多依赖指导教师的辅导，而指导教师对学生而言有一定的局限性。

"6+1"校企联合毕业设计，在开题、中期答辩、末期答辩都邀请不同的专家学者参与，可谓阵容庞大，层面区域范围广泛。这些专家各自的教育背景有很大的差异，他们都有丰富的阅历和自己的个性，对问题都有自己独到的见解，为学生进行方案点评，提供丰富的信息，合理的建议、富有哲理的启发，对学生而言是极难得的学习交流机会，这些特聘专家都是在业界有影响力的学

an opportunity with which other graduation designs cannot compare, it expanded the horizons of the students and their scope of knowledge.

4. Communication mechanism and guarantee of achievements
It is a ubiquity in universities for teachers to be indifferent towards the students' graduation design; usually they do not show enough devotion, which often leads to a poor result. While in this activity, it is specified to carry out periodic inter-group discussions, in this way, the participant teachers and students are encouraged, prompting everyone to ensure a sufficient devotion in the project, so that they can present a corresponding work in front of all the schools and experts, the entire schedule and the requirement on quality is a positive restraint and promotion for guiding teachers and students.
If we compare the graduation design of students who took part in this activity for the last three years with other students' work from the same major, the quality of the former is obviously more guaranteed. We see this as an external constraint; the periodic exchange is promising measure.

5. Clarifying our deficiency and self perfection
Students who participated in this activity of our school major in architecture; this was the first time that they have completed an interior design. While guiding the students, I found out that they often have their own ideas on the functional relationships of architectures, organization of plates, forms and the combinational relationships of structures, but when it comes to the specific contents in interior design, they are often confused or even indifferent. They do not know how to understand or how to handle interior design and organizational processing. Their understanding of the space is rather vague, and limited in depth. The real reason is that they don't understand interior design well enough.
We have set up interior design courses in architectural majors, but it wasn't taken seriously by the students, as long as they get the credit. What we learn in interior design subject and the purpose of learning this course is unclear for many of them. The setting of this course hadn't played its due role while the students study architectural design. They did not understand correctly the concept of interior design in architectural design education; they treat interior design as decoration. While interior design is the continuation and deepening of architectural design, the former is a part of the latter. The nature of interior design is to create the internal space of a building, interior design is not the skin of architecture, nor is it the tool of an internal space. The students didn't understand properly the relationship between interior design and architectural design; in consequence, they paid less and less attention on space, and had a weaker and weaker awareness of it, eventually, they ended up with abandoning their pursuit for space and merely attaching importance to the effect of the facade. And in this way, the architectural design lost its most important element, and the students landed up as masters of external walls.
After analyzing the causes, we learnt that in the future classes, we should strengthen the students' comprehension and cognition of interior design, understand adequately its significance. We will guide them to focusing more on architectural space, and increase the proportion of internal parts in the expression of architectural plans. Hence, the content of their interior design can be presented naturally in their architectural work, and it avoids the separation of these two designs. We will ask the students to embed their design and pursuit of the space in designing a building, increase the students' awareness of the space, and avoid them from focusing only the facade and turning a deaf ear to the form and environment of internal space. Moreover, we would like to encourage them in becoming an architect who takes the pursuit of special design as their real target and prevent them from becoming an expert who knows nothing but the external walls of buildings.

As the development of architectural design industry is turning more and more standardized, the requirement for an integrate design of new architectures has been formed, interior design stands no longer on the waiting list of the construction of a building. This requires us to involve interior design in early stages and throughout the entire architectural design process. It is also a basic guarantee for realizing the construction of an environmental friendly building.

At the same time, a proper and reasonable teaching process establishes a concept that integrate the architecture with interior design for the students, during which, the students will be able to learn and grasp basic methods and techniques of integrative design. It will not only enhance their awareness of space, but also lay a good foundation for their future career. A reasonable teaching content and structure will generate a long-term and positive influence and function.

After three years of guidance on graduation design, I have had plenty of reflections on some of the problems in "Interior Design 6+1" school and enterprise joint graduation design. For instance, should we set up an evaluation standard for the activity or not? How do we make sure that the topics are attractive every year, etc? The influence of this activity is expanding; it is a guiding vision for the development of the education of this discipline, how can the "Interior Design 6+1" school and enterprise joint graduation design go further and step on to the level? It is a question that all the participant universities should be focusing and reflecting on.

者。借助于学会的平台，能邀请这些专家学者到课堂为学生讲授相关的知识并提供尽好的支持，这是任何一个普通的毕业设计所无法相比的。这样的联合毕业设计教学极大地拓展了学生的眼界和与知识面。

4. 交流机制，成果保障

现在高校毕业设计普遍存在教师与学生对毕业设计不重视，投入度不够的现象，导致毕业设计成果不佳，但"室内设计6+1"校企联合毕业设计的活动，由于规定阶段性集体交流，对于参与活动的各个院校师生都是一种激励和促进，促使大家都要保证足够的投入，才能确保在各个阶段提交出相应成果面对所有院校与专家，整体进度与质量要求对指导教师与学生都是一种正面的约束与促进。

从我校三届参与联合毕业设计活动的学生毕设成果在本校同专业内横向比较来看，参加"室内设计6+1"校企联合毕业设计活动的毕业设计质量明显有保障，这是一种外在约束，阶段性交流是一种良好的措施。

5. 明析不足，完善自我

华工大参与活动的学生都是建筑学专业的学生，他们都是第一次做专题室内设计。从指导毕业设计过程当中看到一种现象，学生对于建筑的功能关系、平面组织、形式、型体的组合关系有一定想法，可往往一涉及室内空间设计的具体内容，似乎有些模糊不清，甚至淡漠。但他们在室内空间设计与组织处理上不知如何理解如何着手，有些学生对空间的理解也似是而非，从而导致方案设计的目的性模糊，在深度上也有局限性。究其原因主要是对室内设计理解不够。

建筑专业常规都设置了室内设计课程，但很多学生不太重视，得了学分草草了事。室内设计到底是什么，学习这门课程的目的是什么学生根本不清楚也没太多思考。学生在学习建筑设计过程中，室内设计课的设置没有起到应有的作用。在建筑设计教育当中，没有正确理解室内设计概念，把室内设计当作是装饰看待，不知道室内设计是建筑设计的延续与深化，室内设计就是建筑设计的一部分，室内设计的本质就是创造建筑的内部空间环境，不是做表面的一张表皮，更不是做室内的道具。没有正确地了解室内设计与建筑设计的关系，最终导致学生在学习建筑设计的时候，对空间的追求愈来愈少，对空间的意识愈来愈弱，直至放弃对空间的追求进而只重视外立面效果，使建筑设计失去了最重要的东西。学生经过4~5年的系统学习沦为专做外立面的高手。

分析成因，在今后的建筑设计课程教学过程当中，应当加强学生对室内设计的了解与认识，充分理解室内设计的含义。引导学生更多关注建筑空间，在建筑方案表达方面加强室内部分的比例，让室内设计的内容自然体现在建筑设计的过程中，避免二者完全割裂，让学生在建筑设计过程中把对空间的设计与追求在设计过程中充分体现出来，加强学生的空间意识，避免只想外立面而对内部空间的形态与环境置若罔闻。成为以追求空间设计为真正目标的建筑师而不是只会做外立面的高手。

随着建筑设计行业的规范发展，对于新建筑一体化设计的要求已逐步形成，室内设计不再是建筑完工之后的候补，要求室内设计在建筑设计早期就必须介入并贯穿整个设计过程，这是真正实现绿色建筑的基本保证。

正确合理的教学过程同时也会给学生树立建筑与室内设计一体化的概念，并且在此过程中也能学习掌握一体化设计的基本方法与技巧，既增强了空间意识，也为日后的工作打下良好基础。合理的教学内容与结构会产生长远积极的影响与作用。

经历三届毕业设计指导，对"室内设计6+1"校企联合毕业设计的有些问题也在思考，如是否应该建立"室内设计6+1"联合毕设基本评价标准；如何保证每届毕业设计课题都具有吸引力，联合毕业设计的影响力正在扩大，作为对于学科教育发展有导向愿景的"室内设计6+1"校企联合毕业设计，如何可以走得更好、更远和更上一层楼，很多方面是所有参与活动的院校都应该共同关注与思考的。

Brief Discussion on Renovation of Consumption Space of Old Buildings

Old buildings, especially the historic old buildings which occupy a large proportion of western buildings, have generally three trends from the view of modern urban construction and architectural evolvement: first, in the bloom stage of industrialization, the acceleration of urbanization and rapid expansion of urban space caused the mass construction of infrastructure and new buildings, representing the renovation of old city. Next, in the initial stage of post-industrialization, the expansion of the urban space slowed down due to the decline of environment, efficiency and resource while part of urban functions and consumption transferred from downtown to the fringe area, where the expansion, new construction and renovation coexist; namely, the counter-urbanization emerges. Third, in the flowering stage of post-industrialization, the system of resource recycling and energy transformation was valued, referring to the leading position of the preservation, renovation and regeneration of the buildings. Meanwhile, the concept "re-urbanization" that old buildings are the witness of the past time and separate the areas to own their respective features is proposed. Old buildings can not only endow the urban space with cultural characteristics, but also inject the cultural atmosphere into citizens' life, infusing the city with cultural soul.

1. Attitudes toward old buildings
In China, the reason why many old buildings cannot be preserved is mainly because people suppose it would hinder urban development, which isolates the protection of old buildings with economic progress and neglects the historical nature of the buildings. Due to the unbalanced economic development in China and large-scale demolition projects, urban historic buildings disappear sharply while the cultural diversity among the cities is lessening ; the urban residents concentrate around downtown areas, arising the unbalanced urban development and less vitality ; people's understanding of the integration of renovation of old buildings and consumption area requires improvement.
It can be concluded by the renovation process all over the world that along with the promotion of urban civilization degree, the re-utilization of historical value of the old buildings soars as well. The attitude towards the old buildings has changed from demolition of the old buildings to renovation of the old buildings, which can realize its economic value and cultural value. No matter in Eastern world or Western world, both culture and economy is the necessities to human development, indicating that the development of human civilization is integrated with consumer economy. Thus the increase of consumption brings the development of civilization and vice versa.
No matter westerner or easterner, people always give priority to the promotion of local economy for the purpose of change old buildings to the consumption spaces. The renovated buildings combine both the old feature and new fashions, having reserved the initial soul and injected the fresh blood. It has provided us a sample that the flourishing of historical culture and revival of regional economy can be conducted in harmony as well as a issue to be further discussed.

2. Renovation of the consumption space of old buildings
The renovation of the old buildings initiated long time ago, but people did not realize it was a systematic subject. The real systematic theory emerged in Europe firstly, which was proposed by Eugene Emmanuel Viollet-le-Duc (1814–1879). He was aware that such an adaptable measure could be used to protect the architectural remains. Later, in the early 20th century, Alois Riegl (1858–1905) classified the value of the old buildings and he thought that under the evaluation of the historic buildings, the renovation of the old buildings was also a kind of protection to them. Since 1970s, the renovation of adaptable buildings has started to occupy an ascending position in the architecture field. Table 1 reveals the types of old buildings and classification of their functions and corresponding literature to solve respective function problems. Table 2 illustrates the classification of resources on special buildings, through which we can see the proportion of the consumption space occupies in all the old buildings. The renovation of consumption space is an important aspect of the renovation of old buildings.
In 1960s, a revolution on the renovation of old buildings spread over the US. The first was right the utilization of renovated consumption space in old buildings , such as Ghirardelli Square in San Francisco, the first successful sample of renovation of old buildings in US. The factory, through a series of renovation, became a new consumption space with yard and became US national historic landmark in 1982. The protection of old buildings in China commenced from 1950s. In 1961 the list of first key cultural relics was issued and in 1982 Law of the People's Republic of China on the protection of cultural relics was promulgated. Later Standard of green buildings, Norm of energy-saving renovation of existing buildings I were publicized in 2006 where the key point of renovation was adjusted to all the classes of architecture. However, a majority of old building re-utilization in China is limited in the scope "repairing the old as they had been", which seldom adopts the foreign new methods or innovate ideas and forms certain scale and the methodol

浅谈旧建筑改造消费空间的启示与思考

马辉
刘杰
周立军

Ma Hui
Liu Jie
Zhou Lijun

旧建筑,尤其是历史性旧建筑在欧美国家建筑存量中占有很大的比例,从西方现代城市建设的历程和建筑演变看,大体可以分为三种走向:第一,工业化盛期,城市化加速及城市空间大规模扩张,大量兴建基础设施和新建筑,旧区改造兴起;第二,后工业初期,城市扩张因资源、效率及环境的颓势而减缓,城市部分功能和消费从中心向边缘转移,扩建、新建与保留、改造并举,出现"逆城市化"现象;第三,后工业盛期,资源循环利用和能量转换机制受重视,既有建筑的保留、改造和再生成为主导。"再城市化"理念提出,旧建筑是对历史过去时间的一个见证,也是把不同的区域分割出来,让每个区域有自己的特色。旧建筑不仅能赋予空间的文化特征,同时也把这样的文化氛围注入到生活中,为城市注入文化灵魂。

1. 对旧建筑的态度

我国很多旧建筑保留不下,大部分原因是人们认为旧建筑阻碍了城市经济脚步的发展。把旧建筑的保护与经济进程割裂看待。同时也忽略了建筑历史的性质。由于我国经济的发展不平衡,和大量的拆建工程问题导致了如下问题:① 城市中的历史建筑越来越少,每个城市的文化个体性差异越来越小;② 城市人口密度集中在商圈附近,城市经济发展不均衡,城市逐渐缺少活力;③ 人们对于旧建筑改造和消费空间结合的观念有待提高。通过对国内外旧建筑的改造中,可以看出随着城市文明程度的提升,人们对于建筑历史价值的重新利用大大增加。旧建筑观念上的转变,也从完全拆掉旧建筑转变为改建旧建筑,实现其经济价值和文化价值。无论是东方还是西方,文化和经济都是人类文明发展所必要的。人类文明的发展和消费经济是彼此相生的。消费的提升带动了文明的发展,反之人类文明的进步也会促进消费经济的发展。无论东方还是西方人们对于旧建筑变成消费空间的改造目的都是以振兴当地经济为主。改建后的目标都以新旧结合为主,保留了建筑原始的魂魄,注入了新的血液。这为我们弘扬历史文化的同时也提供了振兴区域经济发展的一个接入点,同时也是有待继续深入研究的问题。

2. 旧建筑改造消费空间

旧建筑的再利用在很早就开始了,但是那个时候人们并没有意识到它会是个系统学科,真正的系统理论兴起首先在欧洲,由 Eugene Emmanuel Viollet-le-Duc (1814–1879) 提出。他意识到这种可适应性的建筑改造手段可以用作保护建筑遗迹。在 20 世纪早期 Alois Riegl (1858–1905) 为这些旧建筑的价值进行分类。同时他认为在他对历史建筑物评估的时候,对那些历史建筑的重新再利用,也是对建筑的一种保护。从 20 世纪 70 年代开始可适应建筑更新在建筑领域开始占有越来越重要的地位。表格 1 显示的历史建筑的类型以及功能分类,同时表明不同的文献资源解决不同的功能问题。表格 2 表达的是不同的资源在特殊建筑上的分类。通过表格可以看出消费空间在所有旧建筑类型所占有的比例。针对消费空间的改建,是旧建筑改造的一个重要方面。20 世纪 60 年代在美国还是兴起了一场关于旧建筑改造的革命,首先就是旧建筑改造消费空间的运用。如旧金山的 Ghirardelli Square,同时这也是美国首个旧建筑改造成消费空间成功的案例。巧克力工厂在 20 世纪 60 年代经过一系列的修复,这个工厂变成了一个带有庭院的新兴的消费空间。1982 年时被列为美国国家历史地标。我国在旧建筑的保护始于 20 世纪 50 年代以后,1961 年颁布了首批重点保护文物的名单。1982 年,颁布了《中华人民共和国文物保护法》。2006 年相继推出了《绿色建筑标准》(GB/T 50378-2006),《既有建筑节能改造规范(一)》等,改造的重点也一再调整至建筑的所有门类。我国的旧建筑再利用大多局限在"修旧如旧",对于国外的新兴手法或者思路,应用相对较少,没有形成一定的规模及自身的再利用方法论。

–ogy of re-utilization.

3. Promotion to the renovation of old buildings by new consumption space

The development of emerging consumption industry in recent years has caused to the emergence of consumption space. Meanwhile, such consumption space can be used to attract consumption groups so as to promote the balanced development of the urban economic zones. The governments can renovate the re-utilizable buildings to not only stress the area with local features and reduce the population density in the city centre .but also promote the margin area but also reduce the construction cost.

Short-term consumption:

Pop-up shop formed in the rural recession period caused by transformation of economy and society and the development of electronic technology. In the post period of World War II, a large amount of population moved from the city to suburban area, leading to the insufficiency of working labors in the city. Therefore, the market began to re-orientate and adopt new technologies to match new productivity mode. However in 1970s, the digital time came and the development of electronic technology industry brought new jobs. The shops, working place no longer concentrated in the shopping malls and office buildings. Such decentralized distribution has provided the pop-up shop with a new environment in short-term lease mode.

Long-term consumption:

The renovation of old buildings are conducted from three aspects: users, buildings themselves and policy balance. As for the traditional consumption space, the renovation of such old buildings for the purpose of long-term operation perhaps focus on the matching degree between building themselves and the projects to be operated, such as the public library of arts college renovated from Mount Royal Station and Goll Bossy Youth hostel. Goll Bossy hostel located in the centre of a city in Croatia had been a shopping mall, but it was destroyed by the local guerrilla due to political factors. Later it became a youth hostel where the past escalators and scenery escalators were preserved. However the spaces in the mall have been replaced by a series of separate spaces which own the basic necessities for sleeping, including the bathrooms. Such a small hostel preserves the original features together with adding the youth atmosphere.

4. Mode of renovation of consumption space of old buildings

From the view of consumption period in the renovation of old buildings, the renovation degree of the old buildings is different due to the length of consumption period. The renovation method of the old buildings for the purpose of consumption mainly lies in their change of their features to adapt to the new conditions. For instance, 1.change of functions: the functions of the old buildings are changed based on the original space. Such method preserves the exterior walls of the old buildings and changes the internal facilities of them, which is suitable for the hotels, art galleries. 5. Redistribution of original space: It fits for the renovation of industrial buildings with large space, such as workshops and railway stations. Such method will redistribute the large space to smaller spaces. 3. Renovation of the structure: the original structure of the old buildings is changed to meet the modern need. 4. Expansion of old buildings. Based on the original structure of the old buildings, the old buildings are expanded so as to integrate with new buildings.

For the development of the consumer economy, various consumption publicity modes and renovation modes emerge. Such pop-up shop brings varying impacts to consumers. Meanwhile, due to its varying styles, it is more suitable for different renovation modes. In addition, the benefits it brings and its experimental opening mode are popular among the sellers and brand designers. Pop-up is a win-win state because both its re-utilization of old buildings and consumption culture are the promotion of economy and culture. Compared with the popular pop-up shop mode, the traditional mode of renovation of old buildings still play an important role in it. The renovation mode for pop-up shops is more like a touch-stone for traditional renovation mode. Its initial "trial operation" mode can help find out the local problems and whether the place is suitable for certain long-term business activities. Therefore, it provides certain economic basis and design demand for traditional renovation mode. The traditional renovation mode changes deeper and lasts longer compared to pop-up renovation. But it requires further discussion that application of pop-up renovation mode and traditional renovation mode to consumption space may have disadvantages and whether it is suitable for architectural mode and architectural types for consumption space in China in the long term.

3. 新兴消费空间促进旧建筑改造

新兴消费产业在近几年的发展导致新兴的消费空间出现。同时这样的消费空间也可以用来引导消费人群，促进城市经济区域的平衡发展。政府对于那些可以重新利用的建筑进行修复，不仅能让这个区域更具有当地的城市特色的同时，减少市中心的人口密度，同时也可以促进城市边缘的发展，减少建设费用。

（1）短期消费：快闪店（pop-up shop）形成于经济和社会的转型和电子技术的发展导致的城市衰退时期。第二次世界大战后期，美国大批人口从城市转到郊区，导致城市劳动力缺少。从而导致市场开始重新定位和采用新技术来适应新的生产力模式。然而20世纪70年代数字时代的到来，电子科技新兴产业的发展带来了新的工作机会。商场，办公空间等不再局限于集中在写字楼和大型购物中心。这样分散以短周期租赁方式的小型空间为 pop-up shop 的形成提供了一个新的环境。

（2）长期消费：旧建筑的改造主要从使用者、建筑和政策权衡来对建筑进行改造。对于传统的消费空间也就是以长期经营为主的旧建筑改造可能更注重于建筑本身和要经营的项目的契合程度。例如由 Mount Royal Station 改造成艺术学院的公共图书馆场所。Goll Bossy 青年旅社，位于克罗地亚的市中心，在一开始的时候是个购物中心，由于政治原因后来被游击队给破坏掉了。后来这个购物中心变成了一个青年旅社。曾经购物中心的自动扶梯和观景电梯被保留了下来。然而购物中心的空间被一系列的独立空间所代替，而这一系列的独立空间内部有睡觉的基本必需品，包括洗手间。这个小小的青年旅社在历史形态下保留了城市的历史风貌，增添一份青春气息。

4. 旧建筑改造消费空间的模式

针对消费空间的旧建筑改造从消费的周期来看，由于有长期消费和短期消费，对于旧建筑改造的程度也不相同。基于消费目的的旧建筑改造主要手法是根据旧建筑的特性来改变其性质适应新的条件。如①变换功能，在旧建筑原有空间的基础上，改变其使用功能。这种方法保留了建筑的外部墙体，更新建筑内部的设备设施，适用于旅社，艺术画廊等。②打散原来空间，这种方法是针对相对大型的建筑空间的改造。将大空间重新进行划分，形成若干小空间，这种手法比较适合改造大空间的工业建筑，如厂房，火车站，仓库和废弃码头等。③对结构的改造，改造建筑物原有结构，满足现代的使用需求。④扩建改造，在原有的建筑结构的基础上，对旧建筑进行扩建加建，使新旧建筑成为一体。

针对消费经济的发展，涌现出不同的消费宣传模式和对建筑的改造模式。其中像快闪店这样的概念带给人们的不仅是百变的冲击力，更适用于不同的建筑改造类型。在经济方面，它带来的经济效益以及这种试验性的开店方式也得到了大量商家和品牌设计师的追捧。快闪形式是一种双赢的状态，无论是对历史建筑的再利用还是对于消费文化上，都是对社会经济以及文化的提升。而相对于当今比较流行的快闪店模式，传统的旧建筑改造模式仍然在发挥着重要的作用。快闪店的改造模式更像是传统改造模式前的一块试金石。通过它前期的类似于"试营业"方式对该地的探索，可以发现该地的问题，以及是否适合长期在此布置展示某些商业活动。这样就为传统的改造方式奠定了一系列的经济基础以及设计需求。传统的改造模式相对比快闪的改造更加深入，更加持久，对于建筑的改造程度要求可能也更大。当然这种快闪模式的改建以及传统改建方式应用在消费空间带来的弊端，以及是否长期适合我国消费空间的建筑模式以及建筑类型等，仍然需要更深层次的探讨。

Exploration and Practice of Teaching Research of School and Enterprise Joint Graduation Design

Our university undertook three sessions of "Interior Design 6+1" school and enterprise national joint graduation design based on the teaching characteristic of graduation design course of our school, this activity is hosted by the China Institute of Interior Design, combining seven universities including Tongji University, Harbin Institute of Technology, South China University of Technology, Beijing University of Civil Engineering and Architecture, Nanjing University of the Arts, Xi`an University of Architecture and Technology, and the Zhejiang University of Technology who joined us afterwards, it was carried out with the cooperation of an enterprise. This activity gave the enterprise an opportunity to understand better the teaching in universities and the students. The game type communication method across multiple universities stimulated the students' consciousness of creativity and their ability. Through the process where the teachers and the students exchange with each other, learn from other's strong points and makeup their own deficiencies, the teaching quality and professional level of the our school's Environmental Art Design major of the Art Department will be largely promoted.

Through the communication channel built by the competition platform, we created an opportunity of bi-directional choices for graduates facing the employment problems; we also raised our influence among all the universities with related majors and the design industry, increased self-confidence of the teachers and the students, and moreover, presented the teaching level in interior design of our school and our figure. We have to say that, this activity being held successfully for multiple times, is the reflect on the teaching practice carried out through vigorous and dynamic teaching attitude, and the new teaching mode, which includes multiple ways like having strong practicality and combing interactive and competitive styles during the courses of interior design majors.

In recent years, our interior design major explored actively the teaching research and the teaching reform

1.A teaching mode keeping pace with the time
As Chinese design is gradually connecting with international design, our competitiveness in the international design world has become an important measurement that examines the quality of our training of design talents, the core and the aim of teaching reform should be conducted based on the market and the social needs. Therefore, our teachers are expected to keep up with the latest trend of this industry, update their teaching content in time, and consider the connection with the social needs as the fundamental entry point.
The environmental design major of the school has continually had reforms on teaching practices. The latter stresses on building itself upon the establishment of the students' comprehensive design ability and an open structure of knowledge, improving constantly the existing teaching contents and teaching methods, training actively the students' creativity, initiative in designing and their general ability to solve practical problems; meanwhile, it focuses on the students' capacity to use what they have learnt in classes on actual projects. In the end, we are looking forward to finding a new teaching mode, and train out professional and competent students in design field who adapt to the modern market and the social needs.

2.A teaching mode that is researching constantly
（1）.In class, we lay emphasis on the mode that combines the theory with cases. In addition to theory teaching parts, the teachers need to cite real cases into their courses, pictures of these real cases allow the students to understand directly the knowledge, which makes it easier for them to put the theory to use. Through comparison and analysis of these cases, we are able to find the priority and the solutions to the difficulties in teaching, in this way, the students will be able to finish their design smoothly.
At the same time, we should highlight the priorities in class, utilize the teachers' researches and actual cases of projects, broaden the innovation space for students, encourage them to discuss actively in class, and then through their interaction with the teachers, we could inspire effectively their ability to find, to analyze and to solve the problems independently.

校企联合毕业设计教学研究探索与实践

刘晓军
何方瑶

Liu Xiaojun
He Fangyao

我校基于校毕业设计课程的教学特点三次承办了"室内设计6+1"全国"CIID"校企联合毕业设计活动，该活动以中国建筑学会室内设计分会为依托，联合同济大学、哈尔滨工业大学、华南理工大学、北京建筑大学、南京艺术学院、西安建筑科技大学以及后来增加的浙江理工大学等7所高校，与一家企业共同开展。该活动增进了企业对院校教学、学生的了解。多个院校之间博弈式的交流方式激发了学生的创新意识与能力。通过教师、同学之间互相交流学习、取长补短共同提高的过程，对我校艺术学院环境艺术设计专业的教学质量和专业教学水平具有很大的提高作用。

通过竞赛这个平台搭建的沟通桥梁，为毕业学生所面临的就业创造了双向选择的机会，提高我校在全国兄弟院校及其设计行业中的影响力，提升了师生的自信心，展示了我校在室内设计教学环节中的教学水准和形象。不得不说，校企联合毕业设计活动在我校的多次成功开展，是我校教师在进行室内设计专业教学时，以充满生机和活力的教学态度，以实践性强、结合互动式、博弈式的多途径新型教学模式进行教学实践的效果体现。

近年来，我校室内设计专业积极进行教学研究和教学改革的探索。

1. 与时俱进的教学模式

随着中国设计与国际设计的逐步接轨，在国际设计界的竞争力成为检验我们设计人才培养的重要标尺，教学改革的核心和目标应依据市场与社会需求进行。因此要求教师与时俱进地掌握学科最新动态，及时更新授课内容，以社会需求"对接"为根本切入点。

我校环境艺术设计专业不断进行教学实践改革，强调立足于构架学生的综合设计能力和开放的知识结构，不断完善现有的教学内容和教学方法，积极培养学生在设计中的创造性、主动性和解决实际问题的综合能力；同时注重学生对课堂所学知识在实际工程项目中的综合运用，最终希望寻找出新的教学模式，力图培养出适应现当代市场和社会需要的专业设计人才。

2. 不断探索的教学方法

（1）在进行课堂教学时，注重采用精讲和案例分析相结合的教学方法。课程进行部分精炼理论教授外，需引用大量实际案例融入教学内容中，运用大量的实际案例材料图片使学生能够直观的理解知识点，方便学生的运用。通过案例的对比与解析，寻找教学重点与难点的解决方案，便于学生能够顺畅的完成设计创意。

同时在课堂教学中突出重点，积极引用教师本人的研究成果和实际工程案例，拓宽学生的创新空间，鼓励学生积极开展课堂讨论，通过师生互动，有效启发学生自主发现问题、分析问题和解决问题的能力。

(2). Many courses in interior design set their design content as actual projects; this can train the students' synthetic ability to express, cooperate and organize in the projects, as well as their ability to communicate with the market. In the concrete practice part in teaching, we propose an open interaction in teaching. The students will step out of the campus and enter into the market, the construction sites to understand the latest trend of interior design, so that the effect which connects art to project, theory to practice can be achieved. Meanwhile, we encourage the students to take part in innovative designs, so that their design capacity will be integrated with the actual needs of the market; and guide them in a targeted way to pay attention to tracking spontaneously the development trend of modern technology. We should also focus on the mainline of the content of artistic design, follow the principles to combine the theory with the cultivation of ability, artistic design with engineering technology, buildings with the environment and interior environment with exterior environment, improve incessantly the overall quality of the students.

(3). The graduation design is combined with competitions, the universities joint with institutes of art; we went through stages of design, practice, exercise, discussion between students and teachers about cases, group conference and on-the-spot teaching. At last, the students present their plans at the defense; a consultation in form of discussions allows their oral expression ability and thinking logicality to get full exercise and improvement. Discussions between teachers and students, students and students and enterprises and students trained the students' ability to communicate and to think comprehensively, agilely and creatively; their resilience while handling real projects; and their organizational and cooperative skills while working in a team. We will be able to strengthen the students' understanding of theoretical knowledge through the practice of interactive teaching, and ultimately accomplish the transmission from absorbing theoretical knowledge to improving working ability.

（2）在室内设计多个课程中将设计内容设置为实际工程案例，培养学生对于实际工程的综合表达、合作与组织和市场沟通等综合能力。在具体的实践教学环节中，提倡开放式教学互动，让学生走出校门到市场、工地去了解室内设计的新动向达到艺术与工程实践结合，理论与实践结合的效果。同时，积极组织学生参与创新性设计，使学生的设计能力与市场实际需求得到有机结合，有的放矢地引导他们注重自主追踪现代科技的发展趋势，紧紧围绕艺术设计内容的主线按照理论教学与能力培养相结合，艺术设计与工程技术相结合，建筑与环境相结合，室内环境与室外环境相结合的理念，不断提高学生综合素质。

（3）毕业设计通过与竞赛结合的方式，院校和设计院结合，通过设计环节、实践环节、习题环节、到师生之间讨论相关案例、集体会看、现场教学。最终到答辩展评环节由学生介绍方案，讨论式的辅导使学生的口头表达能力和思维逻辑性得到了充分锻炼和提高。

Practice and thinking of diversified University Enterprise Joint Graduation Design : Insights on "Interior Design 6+1" 2015 University Enterprise Joint Graduation Design

Abstract: On the basis of the previous two years, universities participating in "Interior Design 6+1" 2015 University Enterprise Joint Graduation Design maintained regional and professional diversification, the event was developed around the transformation of old industrial sites. Universities and enterprises, various professions were joint together, the entire graduation design arranged elaborately implement plans, controlled strictly the quality of processes and achieved an excellent result.

Key words: Diversity, topic, thesis opening, joint graduation design

1.Introduction

2015, the "Interior Design 6 + 1" University Enterprise Joint Graduation Design sponsored by CIID(The China Institute of Interior Design) has entered its third year, the title of this year's topic is " Environmental Design of Nanjing Chenguang 1865 Creative Industry Park". It was undertaken by Nanjing University of the Arts, Xi`an University of Architecture and Technology and South China University of Technology, Nanjing Institute of Historical Architecture Culture took the lead in assigning the topics, with the participation of seven universities : Tongji University, South China University of Technology, Harbin Institute of Technology, Xi`an University of Architecture and Technology, Beijing University of Civil Engineering and Architecture, Nanjing University of the Arts and Zhejiang University of Technology. Multiple professions were involved: architecture, interior design, landscape design, exhibition design, art design, industrial design, etc. In view of the graduation designs of undergraduate education in domestic universities, this event carried out a bold reform in teaching modes and in the contents of courses

2.Diversification brought by the trans-regional and trans-professional mode.

2.1Regional diversification

Universities which join the University Enterprise Joint Graduation Design were Tongji University, South China University of Technology, Harbin Institute of Technology, Xi`an University of Architecture and Technology, Beijing University of Civil Engineering and Architecture, Nanjing University of the Arts and Zhejiang University of Technology, seven universities coming from seven completely different regions and cities. From East to West, these cities have different styles and backgrounds, all possessing distinctive cultural characteristics and urban forms, likewise, citizens and designers from these cities represents different ideas and schools of designing in China.

2.2Diversity in majors

The topic "Environmental Design of Nanjing Chenguang 1865 Creative Industry Park" of this session's joint graduation design was proposed by Nanjing Institute of Historical Architecture Culture; in addition, we were fortunate enough to have the personnel of the Industry Park to participate in this activity; this provided our young designers with first-hand materials of the background of the project. This Creative Industry Park used to be the Nanking Arsenal, which was founded by the general governor of Liangjiang area (Jiangnan and Jiangxi province) LI Hongzhang at the time, this industry park is the cradle of the Chinese military industry and ordnance industry.

Given full play to its historic and cultural advantages, the park exploits protectively the old buildings. The park covers an area of 210,000 square meters, with a total construction area of 110,000 square meters, the orientation and goal of this park is to employ cultural and technological creativity as our theme, and build the park into a fashionable, creative and globally well-known cultural industrial base that integrate together culture, creativity, technology and tourism. Graduates from seven universities are students of architectural design, landscape design, interior design, industrial design, exhibition design, art design and other design majors. They carried out the project from three angles:

（1）. Architectural preservation and landscape design, an Environmental Design of Nanjing Chenguang 1865 Creative Industry Park built on the protection and renewal of industrial architecture heritages in historical cities, combined with A1 and A2 buildings. （2）. Architectural preservation and interior design, the intention of using a multi-functional space that contains armory museum, experimental theater, tourism together comprehensive service center transformed from A1 and A2 buildings, as well as interface design, luminous environment, exhibition design, public art design, etc. （3）. Facility products and guiding system design, environmental facilities in the Nanjing Chenguang 1865 Creative Industry Park and product designs, guiding system designs, cultural diffusion designs, etc. Through an actual project like this, we were able to do effectively a collaborative design of architectural design, landscape design, interior design, exhibition design, industrial design and art design, which enriches the content in a design.

多元化校企联合毕业设计实践与思考
——2015年"室内设计6+1"校企联合毕业设计感悟

摘要：2015年"室内设计6+1"校企联合毕业设计在前两年的基础上，继续保持参加高校区域多元化、专业多元化的特点，围绕旧工业遗址改造展开，学校与企业联合，不同专业联合，整个毕业设计环节精心安排实施方案，严格把控过程质量，取得了良好的效果。

关键词：多元化，选题，开题，联合毕业设计

朱宁克
杨琳

Zhu Ningke
Yang Lin

1. 引言

2015年，由中国建筑学会室内设计分会（CIID）主办的"室内设计6+1"校企联合毕业设计已经进入第三个年头，题目为"南京晨光1865创意产业园环境设计"。由南京艺术学院、西安建筑科技大学、华南理工大学承办，由南京观筑历史建筑文化研究院牵头命题，参加高校有同济大学、华南理工大学、哈尔滨工业大学、西安建筑科技大学、北京建筑大学、南京艺术学院、浙江工业大学7所高校。涉及建筑学、室内设计、景观设计、展示设计、艺术设计、工业设计等多个专业。该次活动针对国内高校的本科生教育的毕业设计环节进行了教学模式与内容的一次大胆改革。

2. 跨地区、跨专业带来的多元化

2.1 区域多元化

本次联合毕业设计的参加高校为同济大学、哈尔滨工业大学、华南理工大学、西安建筑科技大学、北京建筑大学、南京艺术学院、浙江工业大学等7所不同的高校。7所高校可谓来自东西南北中，不同的城市风格迥异，背景不同，都具有鲜明的文化特征和城市内涵，不同的设计师也代表了中国不同的设计思想和流派。

2.2 专业多元化

本次联合毕业设计题目"南京晨光1865创意产业园环境设计"是由南京观筑历史建筑文化研究院命题，更加有幸的是南京晨光1865创意产业园甲方也积极参与到此次活动之中，为年轻的设计师提供了第一手的项目背景资料。南京晨光1865创意产业园，其前身是清末洋务运动期间，时任两江总督的李鸿章于1865年创建的金陵机器局，是中国军事工业和兵器工业的摇篮。

园区充分发挥历史文化优势，保护性开发利用老建筑，园区占地面积21万平方米，总建筑面积11万平方米。园区的定位和目标是：以文化创意和科技创新为主题将园区打造成为国内外知名的融文化、创意、科技、旅游为一体的综合性时尚创意文化产业基地。7所高校的毕业生来自建筑设计、景观设计、室内设计、工业设计、展示设计、艺术设计等多个设计类专业。该项目分为3个角度来进行设计。

建筑保护与景观设计，基于历史城市工业遗产建筑保护与更新，结合A1、A2建筑的南京1865创意产业园景观设计。建筑保护与室内设计，拟将A1、A2建筑改造成兵器博物馆、实验剧场、旅客综合服务中心等功能的内部空间利用与界面设计、光环境设计、展示设计、公共艺术设计等。设施产品与导识设计，南京1865创意产业园环境设施与产品设计、导识（导乘）系统设计、文化传播设计等。通过这样的一个实际课题，有效的将建筑设计——景观设计 - 室内设计 - 展示设计 - 工业设计 - 艺术设计过个专业进行协同设计，即丰富了设计内容，也为同学们在设计、交流阶段提供了不同的视角，全方位的认知自己应该涉猎的相关设计领域，熟练掌握本专业的知识，同时也应该对相关专业有所了解。

3. 联合毕业设计的实施方案

校企联合毕业设计涉及学会、企业及多所高校，参与人员共计约百人，对于活动实施的具体安排变得十分重要，既要多方协调好各单位、人员的时间，也要合理安排活动进程、严控毕业设计质量，最终实现学生提升设计水平、认识

It also offered different angles for the students during their communication and design stage, thus they were able to have a full range understanding of which design field they should dabble in, get hold of the knowledge of their own profession, and what's more the knowledge of other specialties.

3. Embodiment of the joint graduation design

The joint graduation design involved the participation of the Academy, enterprises and numerous schools, with about a hundred participants, the schedule of the activity became rather important. It was necessary not only to coordinate the time of every unit and staff, but also to arrange reasonably the process of the activity, and control strictly the quality of the graduation design. In this way, the students can raise their designing level eventually; perceive the needs and the current situation of the industry. On the other hand, the enterprises get to find out the actual working ability of the students who are going to enter the work force. The entire activity was divided into nine stages: topic research, teaching preparation, opening exploration, design, mid-term examination, profound design, defense review, publication and exhibition for communication. The whole process not only controlled thoroughly the teaching quality of the graduation design, we were also able to exhibit and diffuse the students' design work, personally, I wish that this activity could be spread and shared by more people.

The Joint Graduation Design is an opportunity for students from different universities to exchange their idea; it is also an excellent mode between the universities and enterprises. Throughout the activity, everyone involved had to go through the process of a mental communication and an extension of their vision. If we review the whole process, there are few points we can reflect on.

Firstly: the importance of the topic. The topic of this year "Environmental Design of Nanjing Chenguang 1865 Creative Industry Park" was proposed by Nanjing Institute of Historical Architecture Culture, it is a real project that this institute just completed and it will be continued. The Nanjing Chenguang 1865 Creative Industry Park was reconstructed from the earliest Armory factories of China, but recently, the transformation and utilization of old industrial monuments has become a focus in Chinese society, transformation and utilization of industrial sites projects like the DAD (Dashanzi Art District) in Beijing exist widely in many big cities in China, among which there are plenty of competitive products, whereas there are some projects that made mistakes in its transformation. This year's design was carried on from different angles, including architectural protection, landscape design, interior design and environmental design, which laid foundations for the diversification in the process of the design and the final result.

Secondly: strict control of the teaching quality of the graduation design. During the stage where we proposed and opened a topic, all of the universities needed to communicate fully with each other; we had to take the process of the joint graduation design into consideration while combining the actual situation of every university, specify a detailed and realistic schedule and execute it strictly. The graduation design of this year included nine parts; each one of them had a detailed content and a specific requirement. For important stages such as mid-term examination and defense review, there were separate and detailed requirements, which made it easy to control the quality of the design.

Thirdly: the construction of a cyber teaching management platform. This year, the Joint Graduation Design is one of the sub items on the official website of CIID (China Institute of Interior Design), besides, we set up QQ and Wechat chat rooms, and it allowed the teachers to guide the students in real-time, sometimes even to carry out a trans-regional guidance. In addition, all of the participants were able to learn of the arrangement and the requirements through the internet.

Fourthly: high attention on publicity, and expansion of influences. The trans-regional and inter-professional joint graduation design project lasted six months, its process and content were very diverse, and it was valuable for the teachers and students majoring in relevant specialties and the enterprises. At the beginning of this activity, a variety of advertising tools such as cyber propaganda, book publication and work exhibition were determined. In particular, for the design task of the portfolio of this activity, what we emphasized was not the final result, but the track record of the whole process. For this purpose, we arranged exclusively a graduate to do the design work of the portfolio, to record the designs, photos, sketches, expert forums, what one learnt from the essays, teachers' comments and some others of the entire activity, it had a rich content, and it was intended to present a complete cycle of the graduation design.

The trans-regional and inter-professional school and enterprise joint graduation design improved the students' design capabilities as a group, strengthened their comprehension on design task assigned by the enterprises, promoted mutual understandings of all the young people across the country. In the meantime, it gave the instructors, the topic proposing enterprise and the customer of the project a deeper perception, where everyone benefits greatly.

行业需求与现状的目地，企业也进一步了解了将要走上工作岗位的学生的实际工作能力。活动共分9个阶段展开进行：命题研讨、教学准备、开题踏勘、方案设计、中期检查、深化设计、答辩评审、编辑出版、展览交流。整个过程不但严控毕业设计教学质量，更考虑到设计成果的展览与传播，希望把此次的联合毕业设计活动与更多人分享。

4. 成果与思考

联合毕业设计是不同高校师生的交流与碰撞，也是学校与企业的良好的对接模式，在这个过程中，参与其中的每一个人都进行了一次思想的沟通与视野的开拓。回顾整个过程，也有几点思考。

第一，选题的重要性。本次毕业设计题目由南京观筑历史建筑文化研究院提供，为"南京晨光1865创意产业园环境设计"，是该研究院刚刚完成并不断延续的一个真实题目。南京晨光1865创意产业园是在中国最早一批兵工厂的建筑遗址上改造而来，而近年来，关于旧工业遗址的改造利用正逐步成为中国社会的热点问题，以北京的798文化艺术区为代表的工业遗址改造利用项目，在全国各大城市广泛存在，其中不乏精品之作，也存在个别转型失误的案例。本次设计分别从建筑保护、景观设计、室内设计、环境设计等多个角度开展，为多元化的设计过程及成果奠定了基础。

第二，严控毕业设计教学环节的质量关。在选题与开题环节，各个高校要充分沟通，即考虑到联合毕业设计的进程也要结合各个高校的实际情况，指定一个详细的切实可行的进度计划，并且严格执行。本次毕业设计共包含9个环节，各个环节内容详实，要求明确，对于中期、答辩等关键环节，又单独出了详细的任务要求，很好的控制的设计质量。

第三，构建网络教学管理平台。本次联合毕业设计在中国建筑学会室内设计分会网站建有专门的子项，并且建立了QQ交流群、微信交流群等，实现教师对学生的实时指导，甚至是跨地域指导。对于计划安排、任务要求等也可通过网络让所有参与人员及时获悉。

第四，重视宣传，扩大影响。历时半年的跨地区、跨专业的联合毕业设计，其过程和内容都是相当多元化的，对于相关专业的师生以及企业单位都是有参考价值的。在活动开始之初，就确定了网络宣传、图书出版、作品展示等多种宣传手段。尤其对于此次活动作品集的设计任务，强调的绝不是最终的设计成果，而是全过程的跟踪记录，为此，专门安排了一门毕业生进行书籍的设计工作，记录了从选题到答辩全过程的设计，照片、草图、专家论坛、随笔心得、教师评语等等，内容丰富，要呈现一个联合毕业设计的全生命周期。

跨地区、跨专业的校企联合毕业设计提高了同学们共同协作的设计能力，增强了学生对企业设计任务的了解，促进了与全国各地年轻人的互相了解，同时也让指导教师、命题企业、项目甲方有了更深层次的认知，获益良多。

Research on the Current Situation of Exhibition Design in China

Based on the flourishing exhibition activities and promising prospect of exhibition economy, Chinese universities have initiated the upsurge in the establishment of Exhibition specialty. However, due to the vagueness in the basic concepts, content, professional orientation, industrial management and the education of the specialty, it is still under low-level development. Therefore, the understanding of the current situation and the proposal of the solutions are of great importance to the development of exhibition discipline and exhibition industry.

Section 1 Confusing concepts
Concerning the specialty of exhibition art and technology, professional terms like display design, exhibition design and lay-out design would be referred and cause the confusion. Under the general concept of exhibition economy, many people are used to including the display design into the exhibition design, having degraded the display design. Otherwise, the exhibition specialty is nominated "Art and technology" with an indefinite professional target. Therefore, it is necessary to distinguish among the similar concepts.
1.1 Display design
Display design means "to put thing in a place where people can see it easily". Display design integrates the design of information and establishment of the atmosphere, indicating a purposeful creation of "place and information" Currently; there is no occupation of exhibition design in the Chinese national professional certificate system and catalogue of disciplines and specialties issued by Ministry of Education.
1.2 Exhibition design
In a narrow sense, exhibition is the conference, exhibition and other group activities. In Europe, it is called C&E (Convention and Exposition) or M&E (Meeting And Exposition). In the broad sense, exhibition has wide extensions, including conference, exhibition, and incentive travel; namely, it is mentioned as MICE internationally. Exhibition design is concerned with the initial planning and design works of such group activities. In China there is an occupation of exhibition designer while International Convention & Exhibition Industry Association has been established as the subordinate of China Council for the Promotion of International Trade and Convention & Exhibition offices are set in local government.
1.3 Lay-out design.
Lay-out design mainly refers to the layout and display of exhibits. Generally speaking, the lay-out design indicates the planning and layout in small exhibition hall, excluding the museum and festival design. In China there is an occupation of lay-out designer.
Festival design (or ceremony design) means the initial planning and design festival and ceremony. At present in China exists Chinese branch of International Festivals Events Association, National Festival Committee of CUAES, Chinese Culture Promotion Society, Festival Chinese collaboration and similar agencies. In each province there are also similar agencies.

Section 2 Confusing content
Display design is a comprehensive specialty covering wide-range fields. However, because the modern Chinese design discipline is copied from Western counterparts, the research on classification of the display design by Chinese scholars mainly focuses on the introduction and comment on the western literature. And only a few western universities set the discipline of display design; even if they have set the discipline, due to the openness, practicality and research-orientation of the mature settings of their disciplines, it is difficult for Chinese universities to refer to. Currently in China the content of display design is various and vague, which has three trends: confusion, extension and decrease.
(1) Confusion refers to viewing the display design the same as exhibition design.
(2) Extension means that the display design is considered to include anything relevant to show.
(3) Decrease supposes that the display design is limited to certain scope customarily, such as commercial display, museum display design etc. Such phenomenon is common.
Display can be classified in different ways, such as scope, period, place, exhibit type and area. However, it is difficult for the classification foregoing to clarify the border and composition of the display design. Referring to the basic concept, industrial practice and foreign standard, we can classify the display design into exhibition design, museum display design,

我国展示设计现状研究

朱飞
Zhu Fei

依托于各类展示活动的蓬勃开展和会展经济的火热前景，我国高校也逐渐形成了展示设计专业的办学热潮，但由于该专业从基本概念、内容、职业定位、行业管理，到教育的混乱，没有形成良性循环。因此，了解现状，提出对策，对专业的发展和行业的促进具有重要的现实意义。

1. 混乱的概念

谈到会展艺术与技术专业就会涉及到展示设计、会展设计、展览设计、陈列展览设计等一大堆专业的名词，让人不知所云。很多人在会展经济的大帽子下，把展示设计放在一个大会展的概念里，其结果是弱化了展示设计；要么称为"艺术与科技"，专业的指向非常模糊。因此，搞清这些主要的基本概念是十分重要的。（如图1、图2）

1.1 展示设计

展示，就是将事物"清楚地摆出来给人看"，有"显示"的意思。展示设计融合信息的设计和氛围的营造，是有目地的"场所与信息"的创造。目前，我国相关职业认证中没有展示设计的职业，教育部公布的学科目录中也没有展示设计专业。

1.2 会展设计

狭义地看，会展就是会议、展览等集体性活动，在欧洲，被简称为 C&E (Convention and Exposition) 或者 M&E (Meeting And Exposition)。广义地看，会展的外延很广，主要由会议、展览、节事和奖励旅游四部分组成。也即国际上通常提到的 MICE。会展设计就是会议、展览、节事等集体性活动的前期规划与设计工作。我国有会展设计师职业，在国际贸易促进委员会下设有会议展览业协会，很多地方在政府中设有会展业办公室。

1.3 陈列展览设计

展览陈列主要是指展品的布置和陈列。一般意义上说，展览或者展览陈列设计所指的范围狭小，专指小型展馆的策划与布置，不包括博物馆、节庆礼仪设计等。我国有陈列展览设计师职业。

1.4 节庆礼仪设计

节庆礼仪设计（或称为庆典礼仪设计）是节事、庆典、礼仪活动的前期策划与设计。我国目前设有国际节庆协会(IFEA)中国分支、中华民族节庆专业委员会、节庆中华协作体理事会、中华文化促进会等机构，各省市也有节事的组织机构。

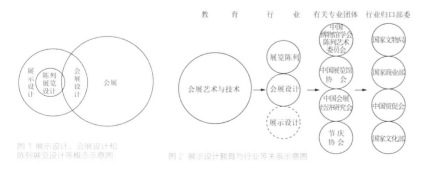

图1 展示设计、会展设计和陈列展览设计等概念示意图

图2 展示设计教育与行业等关系示意图

2. 混乱的内容

展示设计是一门综合性很强、涉及领域广泛的设计学科。但由于中国现代设计学科的移植性，使得学者研究展示设计的分类以西方的引入、评价与译介为主，而国外院校较少设有展示设计专业，即使有，由于其专业的开放性、实践性和研究性，使得我们的院校难以直接参照。目前我国对展示设计所包含的内容，莫衷一是、比较混乱，主要具有3种倾向：混淆化、扩大化和萎缩化。

business display design and festival design.

Section 3 Confusing occupations

In Chinese national occupation catalogues four occupations relevant to the display design are listed, but no such "display design" occupation exists. Lay-out designer is an occupation introduced by original Ministry of Labor and Social Security in 2003, who is engaged in the exhibition research and exhibition design. Exhibition designer is an occupation introduced by original Ministry of Labor and Social Security in 2003, who is engaged in market research on exhibition, project planning (conference, exhibition, promotion display, exhibition facilities) and exhibition operation promotion, interlinking the field of design and economic management. Exhibition image designer is an occupation set by Commercial Catering Service Development &Business skills identification center of State-owned assets Supervision and Administration Commission, who is engaged in exhibition planning and management; China General Chamber of Commerce (affiliated with Ministry of Commerce) also sets an occupation called "lay-out designer". In Lay-out Designer Certificate Industrial Standard issued by Ministry of Commerce, lay-out designer actually refers to furniture lay-out design.From the required professional certificate, knowledge structure, working content, technological qualification and assessment information by the four occupations, it can be concluded that their common base is the display design.

At present the relationship between specialty of design display and its industry is complex. On the relevant national occupation nomination, professional orientation, industrial organization and administrative management exist indefinite factors. As the orientation of the occupation is vague, the integration of the profession and the specialty still has a long way ahead.

Section 4. Confusing education

In 2004, Ministry of Education set the major "Exhibition art and technology" in "Undergraduate major catalogue for Chinese universities". After Science of Arts became a first-grade discipline, it became a major of the "Arts and Technology".

4.1 Various names and contradictory content

Till now, over a hundred universities have set "Arts and Technology" major or similar specialties, among which around 2/3 are featured by exhibition economy and management and the rest are founded on the base of environmental design ,industrial design or visual communication ,focusing on the space and image design, media and exhibition. Even if in professional arts colleges, they will rename the major based on their own understandings and conditions. Now the major has following names : display design, exhibition arts and technology, space display design, exhibition design, exhibition arts, exhibition display design, exhibition planning and design, exhibition planning and creation, exhibition and space design and advertisement& exhibition ,etc. The dazzling names represent the confused understanding of the major.

4.2 Vague cultivation purpose

Only if we have confirmed the requirements for the knowledge, quality and ability of display designers can the education purpose be definite and the universities orientate the major combined with market requirements and respective advantages. Now most of the universities are pursuing to cultivate the students' mould making and aesthetics, planning and organization, literary accomplishment, creative mind, operation ability, modern technology, engineer production, media application, practical experience, communication, economic calculation and personnel management. So many training purposes are described as "cultivating the all-round and application-based talents", representing the fact that all the purposes are vague.

4.3 Deficiency in core courses and system.

At present, many universities just add the courses of other majors and therefore form the major of display design, setting the drawing skill and computer technology as the principal teaching content and focusing on the training of simple skills as well as the simple pile of courses so that the formation of professional knowledge structure and the cultivation feature of the exhibition major are neglected and the core courses of display design are absent.

Under the premise, the prosperous development of Chinese culture and exhibition economy, displaying design meets the golden development opportunity. The solution to the change of confusion of it is that rectification the name of display design with clarifying the basic concepts; defining the discipline scope and integrating the industrial resources; further canceling the national professional certificate of exhibition designer and lay-out designer and setting the display designer to define its national occupation standard and obtain the professional identity in the society. Besides, the relevant majors, design- display industry, relevant academic society and competent administration should be coordinated to gradually manage the industry and discipline unitarily; finally, under the discipline of design, the cultivation purpose of the display design should be defined and the systematic and coherent core course system should be established.

（1）混淆化就是把展示设计等同于会展设计。

（2）扩大化就是认为展示设计包罗万象，所有有关"秀"的成分都属于展示设计。

（3）萎缩化就是习惯性地把展示设计囿于某一部分，如商业陈列，或者博物馆展示设计，等等。这种现象比较普遍。

展示分类的方式众多，如按照规模、时间、场所、展品类型、地域等进行分类等，但这些分类很难探明展示设计的边界和构成关系。结合基本概念、行业特征和国外经验，我们可以把展示设计划分为会展设计、博物馆展示设计、商业展示设计和节庆礼仪设计。

3. 混乱的职业

我国有关展示设计的职业目录中一共有4个职业，但没有展示设计职业。陈列展览设计师是国家劳动和社会保障部2003年推出的职业，是指从事陈列研究、展览设计的专业人员；会展设计师是国家劳动和社会保障部2007年推出的职业，从事会展的市场研发、项目策划（会议、展览、节事活动、促销演示、场馆展具）和运营推广等相关工作，横跨设计类和经济管理类两大学科；会展形象设计师是国务院国资委商业技能鉴定与饮食服务发展中心设置的职业，主要是从事会展策划和管理方面的工作；中国商务部中国商业联合会也确定了一个陈列设计师的职业。在商务部立项签发的《陈列设计师资格条件行业标准》中，陈列设计其实是指家居陈设设计。

从以上4个职业公布的所需专业资格、知识结构、工作内容、职业要求、技术条件，以及考核信息来看，它们的共同基础就是展示设计。

当下的展示设计专业与行业之间的关系比较复杂，在国家职业命名、社会职业定位、社会行业团体组织及行政隶属管理中都存在着分离的不确定因素，职业的定位不清晰，职业和专业的整合之路令人堪忧。

4. 混乱的教育现状

教育部在2004年高等学校本科专业目录中设置了"会展艺术与技术专业"，艺术升为门类学科以后，目前成为"艺术与科技"学科中的专业。

4.1 名称繁多、内涵矛盾

目前，全国已有超过百所高等院校设立了"艺术与科技"相关专业或专业方向，在这些院校中，约有2/3是以会展经济和管理为重点的类型；其余的院校是在以原环艺设计、工业设计，或者视觉传达设计基础上开设的，侧重于空间、造型设计，以及媒体、布展等内容。即使是设计类专业院系，也会按照自己的情况和理解给该专业再以命名。现专业称谓有展示设计、会展艺术与技术、空间展示设计、会展设计、会展艺术、会展展示设计、会展策划与设计、会展企划与创意、会展与空间设计、广告与会展等。名称的不同反映出理解的混乱。

4.2 培养目标不清晰

我们只有明确了展示设计人才所要具备的知识、素质和能力要求，才有可能确立教学目标，并结合市场的实际需求和学校的办学优势与特色进行专业定位。目前大多院校都追求培养学生造型审美、策划组织、文学素养、创意思维、动手能力、现代技术、工程制作、媒体应用、实践经验、沟通交流、经济核算、人员管理能力等。什么都想要，美其名曰是培养复合型、应用型的人才，其实是职业、专业、人才目标不明确。

4.3 缺乏核心课程、系统性不强

目前，许多院校只是将自己原有的其他专业课程简单相加，把绘画技能、电脑技术等成为教学的主体，偏重于简单技艺的培养，课程之间的简单叠加，忽视了专业知识结构的形成和展示设计专业特色的培养，缺乏展示设计的核心课程。

在我国文化大繁荣和会展经济大发展的前提下，展示设计迎来了难得的发展机遇。如何改变目前混乱的现状，这就需要做到：理清基本概念，为展示设计正名；明确专业范围，整合行业资源；取消会展设计师和陈列展览师等各类国家级相关职业，设置展示设计师，明确其国家职业标准，取得社会的职业认同；另外还要统一协调专业、行业和相关学术团体，政府有关部门应逐步统一管理展示设计职业和学科；最后，在设计学科下，明确展示设计专业的培养目标，建立连贯的核心课程体系。

Discussion on the "whole-process node-control" Teaching Design to Graduation Design

Graduation design is an important stage in the university cultivation system. For this stage, different universities have respective organization modes. After years of educational practice, we propose application of the "whole-process node-control" teaching design and management mode to the graduation design. Graduation design, as the last education phase in the university, requiring students to use all the grasped knowledge comprehensively so as to expand and deepen basic knowledge and professional knowledge and improve the research and independent working ability. "Whole-process node-control" education design reinforces the management and direction to the graduation design, which has positive effect on increasing the quality of graduation design and understanding of basic design process and method by the students.

1.Flow-management education purpose
Graduation design is a key component of undergraduate student cultivation plan, also the last learning stage of students on campus, as well as the examination of whole talent cultivation system and cultivation quality of the university. Department of Environment Design with Zhe Jiang University of Technology requires the graduating students to utilize the mastered basic theory, basic knowledge and basic skill (e.g. drawing, CAD, translation, literature review) systematically apprehensively, with independent analysis, so as to master the preliminary forming method and skill of research-based design scheme. In the instruction to the graduation design, the node-control is emphasized in the whole process: the teaching purposes are divided into practicable and controllable periodical targets whose accomplishment can form systematic and formal constant-quality graduation design. Along with the periodical graduation design, the students have reinforced the comprehensive utilization of their professional knowledge and increased their professional quality and design ability.

2. "Whole-process node-control" teaching method
The teaching method of graduation design ,regarding the research-based study as the main thread and emphasizing the whole-process node control, focuses on the logic and systematic ness of the research process and the pace and integrity of the process management. In the experimental teaching practice of CIID interior-design 6+1 Enterprise& University graduation design, the general teaching procedures consist of selection of subject, opening report, research, analysis, concept definition, content deepening, scheme formation, mid-term inspection, finalization and evaluation(thesis defense, exhibition and contest),etc. The orderly initiation and accomplishment of the phases are the premise and guarantee of the high-quality graduation design of which the opening report, mid-term inspection and thesis defense are the key nodes.

Opening report is the necessity of the graduation design and high-quality opening report is the foremost guarantee of the outstanding accomplishment of the graduation design. During the preparation for the opening report, the students must carry out site investigation and consult a large amount of relevant materials, including background information and design theory. Such self-learning in the phase can motivate the students' agency and cultivate the students' learning customs while the students can have preliminary understanding of the general design work at the initiation period and pre-judgment to the key points and difficulties. In the students' reporting process of CIID interior-design 6+1 Enterprise and University joint graduation design, they illustrated the information of sorted sites and similar samples and had face-to-face communication with the experts, having prepared the basic for design and formed basic concepts and thinking.

Mid-term inspection requires each student to report his/her research progress and answer the mentor group questions. Based on the student's understandings of the subject content and requirement, the mentors check if the design content is consistent with the subject ,the overall design scheme is reasonable , it can meet the function requirement and own aesthetic perception or not. In the mid-term inspection process of CIID interior-design 6+1 Enterprise-University graduation design, the expert group and mentor group commented on the students' schemes seriously and gave them detailed instructions so that the students could review their design scheme from multi-perspectives and therefore break through own design bottleneck ,having laid the solid foundation for the high-quality work.

The thesis-defense phase is the final of the graduation design, indicating the evaluation and feedback phase of the whole teaching process to the graduation design rather than the completeness of the teaching.

对"全过程节点把控"式毕业设计教学设计的探讨

吕勤智
黄焱

Lv Qinzhi
Huang Yan

毕业设计是人才培养体系中的重要阶段,针对这一环节的教学设计不同学校有各自的组织方式。在多年的教学实践中主张毕业设计的"全过程节点把控"式的教学设计和管理模式。毕业设计作为大学最后阶段的教学环节要求学生综合运用所学知识,进一步扩展和深化基础知识和专业知识,提高研究能力和独立工作的能力。"全过程节点把控"式的教学设计强化了毕业设计过程管理与指导工作,对提高毕业设计教学质量和学生掌握基本的设计程序和设计方法,提高设计水平起到积极作用。

1. 过程化管理的教学目标

毕业设计是本科人才培养方案的重要组成部分,也是学生在校学习的最后环节,同时也是对整个人才培养体系和培养质量的检验。浙江工业大学环境设计学科针对毕业设计教学目标设定为使学生在毕业设计中获得能力的提升,在这一环节要求学生综合、系统、全面地运用所学基本理论、基本知识、基本技能(如绘图、计算机辅助设计、翻译、查阅文献等)和独立分析、解决问题的能力,初步掌握研究型设计方案生成的方法和技能。毕业设计教学中强调全过程节点把控,通过将教学目标分解成可落、实可管控的阶段性目标,以阶段性节点目标的完成,形成体系化、规范化质量恒定的毕业设计。学生在分解的阶段性毕业设计环节中,不断强化专业知识的综合性运用,提高了专业素养和设计能力。

2. 全过程节点把控的教学方法

我校毕业设计的教学方法以研究性学习为主线,强调全过程节点把控,注重研究过程的逻辑与系统性,注重过程管理的节奏与整体性。在参加CIID"室内设计6+1"校企联合毕业设计的实验性教学实践中,整个毕业设计教学步骤包括选题、开题、调研、分析、概念、深化、方案、中检、表达、评价(答辩、展览、竞赛)等,这些环节的有序开展和完成是高质量毕业设计的前提和保障。其中开题、中检、和答辩3个环节是毕业设计的重要节点。

开题环节是做好毕业设计不可或缺的一环,进行高质量的开题是出色完成整个毕业设计的首要保证。学生在准备开题的过程中需要进行场地勘察和查阅大量与课题相关的资料,包括背景资料及相关设计理论。通过这一过程中的自学环节,将充分调动学生的能动性,培养学生的学习习惯。同时使得学生在设计的开始阶段即对整个设计工作有一个初步的了解,对毕业设计中的重点难点有一个预先的判读。而在本届CIID"室内设计6+1"校企联合毕业设计开题汇报阶段,学生将梳理过的场地资料、相似案例一一展开,与专家进行面对面的沟通,为设计做好了基础准备工作,形成了基本思路及初始概念。

中期检查时每位学生需汇报进展情况,回答指导教师(组)提出的问题。根据学生对课题内容与要求的理解,检查毕业设计的内容与题目是否一致,设计方案的整体是否合理,能否满足功能并具有形式美感。CIID"室内设计6+1"校企联合毕业设计中期检查阶段,专家组及导师组对每组学生方案进行了认真点评与具体指导,让学生能够从多视角重新审视设计方案,突破自身设计思维瓶颈,为最终高质量的完成毕业设计奠定了扎实基础。

毕业设计终期答辩环节是整个毕业设计工作的最后环节,但并不意味着毕业设计教学过程的结束,而是对整个毕业设计教学过程的评价与反馈阶段。通过答辩可以检验学生毕业设计的质量,考查指导教师的指导水平,锻炼学生归纳、总结设计成果的能力,考核学生的逻辑思维、语言表达以及PPT制作等多方面的素质。在本届CIID"室内设计6+1"校企联合毕业设计终期答辩阶段,专家组及导师组对每组学生就方案本身及所表达的设计理念、呈现的设计手法及相关知识进行了提问,对学生方案及综合素质做出了综合评定。在毕业设计的完成

The thesis defense can verify the quality of students' design works, evaluate their logical thinking, language performance and PPT-production as well as inspect the mentors' instruction; more importantly, it can improve student's abilities of summarizing their design works. In the thesis-defense process of CIID interior-design 6+1 Enterprise and University joint graduation design, the expert group and mentor group raised the questions to the students about the design schemes themselves and their represented design theories, design methods and relevant knowledge upon which the design schemes and overall quality of the students were assessed. The accomplishment phase of the graduation design helped both the students reevaluate their design scheme and the teachers notice the key-focus in the instruction process.

The three above phase are closely-interlinked parts as well as dependent bodies. The students should comprehend and master the relevant theories and logic system in each phase and compensate their weaknesses pointed by the experts and mentors through the instruction or self-learning so as to improve the growth of self design ability. Meanwhile, the mentors from time to time should be aware of leading the students to approaching a definite and efficient design process with which the students' design abilities can be improved and acquainted.

3. Whole-process management

Whole-process management has two meanings. First, it refers to the management of mentors. The mentor, as the teacher who contacts students most frequently and gives them most instruction and help, demands high motivation because it is a key to the graduation design. Only if the mentor strictly manage the students and arrange the working plan can the students devote sufficient time and energy to the graduation design. The setting of opening report, mid-term examination, thesis defense and other sub-phases set the periodical detailed requirements for the mentors on the instruction to the students and the basic control on the design progress. The inspection to students also concentrate in opening report (early period), mid-term inspection (mid-term period) and thesis defense (final period).The early period focuses on the inspection of selection of topics and site analysis; the mid-term inspection lie in the progress check, scheme concept and overall arrangement; the final-term inspection targets at the quality, depth of the design and preparation for the defense. In addition, the mentor is required to have the discussion on the design scheme with his/her student at least once a week, including the overall planning, rationality and details of the design. In this way can the student improve own ability and bestow the teachers with pleasure and respect by that both teachers and students make progress by learning from each other. Furthermore, the students are required to write work logs recording the impression, inspiration, difficulties and understandings in the design process, which can help sort out the design logic and provide the basic materials for later composition.

4. Conclusion

The participation in experimental teaching activity of CIID interior-design 6+1 Enterprise and University joint graduation design has enhanced the academic communication the department of environmental design with Zhe Jiang University of Technology with domestic counterparts and promoted the mutual understandings. It has also reinforced the management and instruction to the graduation design, improved the teaching quality of graduation design and students' design abilities and further promoted relevant teaching research and the construction and development of the discipline of design. Also it has played an important role in the improvement the teaching quality and development to the high-level of specialty of environment design of Zhe Jiang University of Technology.

阶段赋予了学生评价自身方案的新意义，同时给予了教师在指导过程应予重点关注内容的启迪。

以上3个环节既是环环紧扣的阶段工作，同时也是相互依存，有机联系的整体。学生需将每个阶段所涉及的相关理论、逻辑体系、技术要点融会贯通，对专家和老师指出的不足之处通过教师指导或自学等途径来弥补加强，增强自身设计能力的成长性。而指导教师也需在整个过程中时刻注意引导学生清晰高效的推进设计过程，并从中掌握和提高设计能力。

3. 过程管理的全过程化

全过程化的管理包含两方面的含义。首先，是针对毕业设计指导教师的管理，作为与学生接触最多、给予学生指导与帮助最大的指导教师，如何调动其积极性，是保证本科毕业设计关键之一。只有指导教师对学生严格管理、合理地安排工作计划，才能切实保证学生投入充足的时间与精力完成毕业设计工作。而开题。中期检查、终期答辩这三大环节和其他子环节的设置，使得教师在毕业设计指导上有了更为明确的阶段性要求，在进度安排上也有了基本控制。在对学生阶段性检查中，着重在开题（前期）、中期检查（中期）、终期答辩（末期）3个阶段。前期重点检查设计选题，场地分析等内容；中期重点检查设计进度、方案概念、整体布局等内容；后期重点检查毕业设计的质量、深度和答辩情况等内容。同时要求师生间开展每周不少于1次的方案讨论会，对方案的整体构思、合理性、细节等进行经常性的讨论。在促成学生能力提升的同时，也给予了教师在学生发现问题、解决问题的成长过程中感受到的教学相长的乐趣与尊严。此外要求学生撰写工作日志，记录在毕业设计环节的所思所学所悟，记录随时迸发的灵感、难点及其他点点滴滴，有助于梳理设计逻辑，也为未来文本的制作，论文的书写积累基础性资料。

4. 结语

经过参与CIID"室内设计6+1"校企联合毕业设计实验教学活动，加强了浙江工业大学环境设计学科与国内相关学科和兄弟院校间的学术交流，增进了相互了解。同时也强化了毕业设计过程管理与指导工作，对提高毕业设计教学质量和学生设计水平起到积极作用，推动了教学研究和设计学学科的建设与发展。对于推动浙江工业大学环境设计专业教育质量提高和学科向高水平方向发展起到了重要作用。

CIID "室内设计 6+1" 2015（第三届）校企联合毕业设计
CIID "Interior Design 6+1" 2015(Third Scssion)University and Enterprise Joint in Graduation Design

活动印记
Activity Mark

Activity mark 活动印记

寄 语 Message

陈卫新
CIID 理事
南京观筑历史建筑文化研究院院长

邹超
南京晨光 1865 公司总经理

6+1 校企合作的教育形式，应该说非常重要。从学生的角度来说，既能实际接触到一些设计最新的发展趋势，又能够关注到传统本身的一些东西，给了同学们一个很好的拓展自己专业素养的平台。这是"室内设计6+1"的一个特点。

未来的路很长，祝福此项活动能更加坚实地走下去。应当说现在基础非常好，机会也非常好，但这些都不能代替自己的努力，这一点可能是同学们要有思想准备的，要面临很多的挫折，另外也要不断地调整自己、充实自己。从小事做起，会有自己的天地。

关注社会，用自己的设计语言，真诚的面对设计，不必盲从流行与玄虚，独立的思维与成熟的品格是走向成功的基础。

The education form of 6+1 college-enterprise cooperation is very important. The students can practically contact with the latest development trend of design and also focus on some traditional things. It gives a good platform for students to expand their professional accomplishments well. This is a characteristic of 6+1.

The road in future is very long. I wish this kind of activity can be held more consistently. The present basis is very good and the chance is good too. But, all of these cannot replace one's own hard-working. This point should draw attention of students and the students should prepare to receive defeats and continuously adjust one and enrich one by doing small things at the beginning and would have one's own world.

Pay attention to society and use one's own design language to face the design. Don't follow blindly the fashion and hollowness. The independent thinking and mature personality are basis to success.

"战以止战、兵以弥兵，制造正义的剑是为了保卫和平"，此一精神延续至今，而航天科工新的产品格局也一直捍卫并传承这一精神宗旨。2007年，顺应结构转型、退二进三的经济形势转变，1865作为活的军工建筑化石，再次投入使用，并被赋予新的使命，按照"文化+"的定位，"文化+创意""文化+设计""文化+科技""文化+旅游"的思路，打造城市文化综合体，获得了"国家级重点文化产业试验园区""江苏省文化产业示范基地""江苏省高品质城市生活空间"等一系列称号。

我们有幸迎来了中国建筑学会室内设计分会主办的此次活动，在继东南大学、南京大学、美国南加州大学建筑学院、澳大利亚新南威尔士大学的专项研究之后，又有一批建筑设计专业、景观设计专业的行业翘楚，对1865进行专项研究。

希望1865独有的建筑特色能给在座的每一位，提供一次美妙的观感对话和设计体验，能够丰富每一位的学习研究历程。我想同学们的作品对我们也多有启发，在老师的打磨下定能用于实践，这也是协会的宗旨之一，也是产学研的目标。齐聚如此众多的国内重点高校专业人士，对金陵制造局开展研究，是我们的荣幸，也希望各位持续关注1865的后期发展，并诚挚邀请各位可以参与到1865的发展中来。

The spirit "fighting to stop the war and having soldiers to make up the soldiers, manufacturing justice sword to protect the peace" continues to the present day and is defended and inherited by new product pattern of Aerospace Science and Industry. In 2007, to conform to structure transformation and transformation of the economic situation of retreat into three, 1865 as living military industry building fossil is reused and imparted by new mission. In accordance with the position of "culture+", and concepts of "culture +creation", "culture +design", "culture+ science and technology", "culture +tourism", we create urban cultural syntheses and thereby obtain a series of titles such as "National key cultural industry test zone", "cultural industry demonstration base of Jiangsu province" and "Jiangsu province high quality city living space".

It is fortunate that the activity is hosted by interior design branch of China construction institute. Following the special project researches of Southeast University, Nanjing University, School of Architecture of University of Southern California, and University of New South Wales of Australia, another batch of industry outstanding persons of architectural design profession and landscape design profession conduct the specific project research on 1865.

I hope that the unique building characteristics of 1865 can provide splendid view feel conversation and design experience for every one and can enrich studying and research course of every one. The students' works give us great inspiration and will be certainly used for practice by polishing by teachers, which is also one of aims of the institute and an object of industry-university-research. Many professional persons from the domestic key universities are gathered here to study the Jinling manufacturing bureau, this is our honor, so we hope all of you can continuously focus on the future development of 1865 and we sincerely invite you to participate in the development of 1865.

左琰
CIID 理事
同济大学建筑系教授

三个多月的毕业设计短暂而又忙碌，却给人留下了深刻印象，值得回味。一是选题好，由南京观筑历史建筑文化研究院陈卫新院长出题，将前身为晚清李鸿章创办的金陵制造局的南京 1865 文化创意园区为设计对象，聚焦园内两个巨大的锯齿状屋顶建筑进行保护性改造利用。作为中国四大兵工厂之一，南京 1865 园区从命运多变的厂史到各时期建造的不同建筑样式都具有其极为特殊的历史价值和时代特征，而目前产权人晨光集团致力于推动国防科技力量，也自 20 世纪 90 年代初至今为国内外著名佛教圣地铸造各类大型佛像，为佛教事业的繁荣做出了重要贡献，因此从园区原先的兵器制造转向止杀向善的大佛铸造，很好地诠释了战争与和平的对立与统一关系，其遗产价值已超越了一般工业遗产地的价值和意义。二是有竞争，由于此次的活动规则较以往有所调整，限定每校只有两组进入最后评选，而我们学校分为 3 组，这意味着有一组同学会无缘最终参评汇报，因此师生的竞争意识和挑战意识也相应加强，全过程必须全身投入才能确保较好的发挥和指导。三是平台宽，不同学校的师资背景、教学理念和学科方向都不相同，每年的命题单位和评委组成也不相同，此次设计选题分为三大具体方向，以工业遗产保护为前提，分别侧重室内设计、景观设计和展示设计 3 个领域，参与此活动的师生可按照统发的框架任务书进行内容和版式的调整，通过开题、中期到最后答辩 3 次汇报互相学习和交流，大大开拓了教与学的视野。

祝愿室内设计 6+1 联合毕业设计活动能保持这种势态，越办越好！

More than three months of graduation design is short but busy. It left a deep impression to the person and is worthy of aftertaste: first, the topic is very good. The topic is assigned by a director Chen Weixin of Nanjing Guanzhu Historical Building Culture Research Institute, taking Nanjing 1865 culture creative parks of which the predecessor is Jinling Manufacturing Bureau established by Li Hongzhang in late Qing Dynasty as the design object and focusing on two huge zigzag roof buildings to conduct protective transformation and utilization. As one of the Chinese four arsenals, Nanjing 1865 parks have extremely special historical value and characteristics of the times from factory history of various fates to different building styles built at different periods. At present, the property owner Chenguang group devotes itself to push national defense science and technology strength; also, it casts many large-scale figures of Buddha for famous Buddhist holy land at home and abroad from the early 1990s to now and makes important contribution to prosperity of Buddhism. Thus, the diversion from the original weapons manufacturing to Buddha casting which represents stopping killing and doing good turns of the parks interpret the opposite and unified relationship between war and peace well. The heritage value thereof exceeds the value and meaning of general industrial heritage. Second, it is competitive. Activity rules at this time are adjusted compared to the previous activities, it is defined that only two groups of each college can enter into the final vote. But, there are three groups of our college, which means that one group of students, will not be allowed to participate in the final comments and report. As such, the sense of competition and the sense of challenge of teachers and students are strengthened correspondingly. They should participate in the whole process from head to foot to thereby ensure good performance and guidance. Third, the platform is wide. The teacher background, teaching concept and discipline orientation of different colleges are different. The unit to assign the topic and the composition of judge are different too. At this time, the topic includes three specific orientations, taking industrial heritage protection as the premise to respectively focus on interior design, landscape design and exhibition design. Teachers and students participating in this activity can adjust the content and format in accordance with the issued frame task, study from each other and communicate with each other through three reports: topic proposal, mid-phase and final defense, which greatly opens up the view of teaching and studying.

I wish the interior design 6+1 joint graduation design activity can keep such an attitude and do better in the future.

金戈
南京地方史志专家

建筑，尤其是历史建筑，往往与重大事件相关联。比如，1865创意产业园区的锯齿形多跨厂房，既是中国近现代工业建筑的代表之作，更是抗日功勋李承干先生率领他的团队，对清末金陵制造局实施彻底改造的产物。

环境设计离不开对建筑及其历史的认知。由此，我们不仅要认识到，他们在建造锯齿形厂房时对新材料与新技术的纯熟运用；更要体会到，在民族面临危亡之时，他们建立在相互信任基础上的，共同实现远大抱负的精神追求。

我希望，作为有志于美化生活的设计者，应该从前人及其伟大作品那里，获得营养，获得启迪。

Buildings, especially historical buildings, are generally associated with important events. For example, the zigzag multispan frames factory buildings of 1865 creative industry parks are not only representatives of China's modern industrial buildings but also products of complete reconstruction of Jinling manufacturing bureau in the late Qing Dynasty by anti-Japanese hero Li Chenggan and his team.

The environment design cannot be done without perception to the building and its history. Thereby, we should not only realize the skillful application of new materials and new technologies by them when constructing zigzag factory buildings, but also perceive their spirit pursuit for realizing the ambitious aspiration together on the basis of mutual trust when the nation's existence is in peril.

I hope that the designers that aim at beautifying the life should obtain nutrition from the predecessors and their great works and get enlightenment.

CIID"室内设计6+1"校企联合毕业设计，经过3年的实践，这种毕业设计的新形式，在导师和同学们的共同努力下，逐步完善和成熟，提高了同学们参与的积极性，导师们也越来越有信心。

这种形式的毕业设计的最大特点，是它具有极强的开放性。它不仅最大程度上保持和发挥本校的专业教学特色，又可以通过开题、中期答辩、毕业答辩的机会，观摩和学习到不同专业背景的其他的院校专业知识，可以扩大同学们的视野和专业知识面，对今后的工作会有很大的帮助。与此同时，由于与设计企业的接触，使同学们更多的了解到实际设计项目的要求。

校企联合毕业设计，不仅是有关院校的参加，更重要的是要有一个设计企业的全程配合，和一个适合的命题，这个题目要是实际设计项目，难度适当，适合同学们的毕业创作，由于时间限制，设计的规模要适当，三届的毕业设计企业都给与这项活动极大的支持。

最后，校企联合毕业设计是开放式的进行，不是局限在一地一校之内可以完成，它的一个周期有命题、开题、调研、专业培训、中期答辩、直到毕业答辩几个不同的环节，往往会涉及到几个城市和院校要各校领导给与了很大的支持。

CIIID "interior design 6+1" college-enterprise cooperation graduation design has been practiced for three years. Such a new form of graduation design is gradually perfect and mature under the joint effort of mentors and students. It improves students' enthusiasm to participate in and the mentors are more confident.

The biggest feature of this kind of graduation design is the very strong openness. It keeps and plays the professional teaching feature of the own college at the greatest degree. From it, the students can view and study professional knowledge from the other different professional background colleges through the chance of topic proposal, mid-phase and final defense, which greatly opens up the view of teaching and studying. This can enlarge the students' view and professional knowledge and help working in the future. At the same time, in view of contact with the design enterprise, the students learn about more requirements of actual design projects.

The college-enterprise cooperation graduation design requires participation of the related colleges and more important the whole course cooperation of one design enterprise and a suitable topic. This topic should be the actual design project and has a proper difficulty and is suitable for the students' graduation creation. In view of time limit, the scale of the design should be proper. The graduation design enterprises of three years all give very big support to this project.

Finally, the college-enterprise cooperation graduation design has an open style and cannot be finished by being limited in one land and one college. One of its periods includes several different steps: assigning a topic, topic proposal, research, professional training, mid-phase defense and graduation defense. This often relates to several cities and colleges and needs great support from leaders in the respective colleges.

李书才
CIID 资深顾问
CIID 竞赛委员会主任

　　毕业设计作为一个相对长时间上进行的课程教学，从选题、开题、构思、到最后毕业设计的展览展示，是对学生4年学习的全面考核，是别的课程所不能比拟的。

　　"室内设计6+1"是6所院校加1个社会实体，这是院校与社会对接的教学机制，是院校教学的开放平台。在这个阶段里，需要学生综合性地运用本科所学的各项专业知识去发现、分析、系统解决社会问题的综合能力。使学生的综合素质、专业技能，还有职业素养得到进一步的提升与整合。换句话讲，毕业设计是学生从学校走向社会的一个过渡阶段，而笔者所观察的"室内设计6+1"的校企联合教学活动也是希望通过毕业设计培养学生的创造性思维和实际应用能力，真正做到让学生理论联系实践。

　　毕业设计作为课时最长、最具实践特征的教学环节，选题是这一环节中的关键。这一阶段付出的精力不应少于后续的方案设计阶段。俗话说："题好文一半"，好的选题能获得事半功倍的效果。今年的"CIID室内设计6+1"的设计选题是以南京晨光1865创意产业园的改造为题，毕业设计选题的目的不仅仅是让学生完成设计方案，更是结合社会有积极意义的项目意义在于让学生明白设计的目的，同时进一步探索设计的社会价值。

林学明
CIID 副理事长
集美组设计机构创意总监

Graduation design, as a course teaching conducted relatively for a long time, is an overall assessment of students' four-year studying from choosing a topic, proposing a topic, concept to the final exhibition of the graduation design, which cannot be compared with the other courses. The "interior design 6+1" includes six colleges plus one society entity. This is the teaching mechanism where the college is joining with the society and is also an open platform of college teaching. In this phase, the students should apply the professional knowledge learnt from undergraduate courses to realize, analyze and systematically solve the society problem, so that the students' comprehensive quality, professional skills and professional quality are further lifted and integrated. In other words, the graduation design is a transitional phase where the students walk from college to society, and the college-enterprise cooperation teaching activity of "interior design 6+1" viewed by the writer also hopes to train students' creative thinking and actual application ability through the graduation design to really make the students' theory in association with practice.

The graduation design is the teaching step having the most class hours and practice characteristics, wherein choosing a topic is the key. The effort paid in this stage should not be less than the subsequent scheme design stage. As the saying goes: "a good topic is a half of the article", which means that the good topic can obtain the effect of yielding twice the result with half the effort. The design topic "CIID interior design 6+1" of this year takes transformation of Nanjing Cheguang 1865 creative industry parks as the theme. The object of topic of the graduation design is not to only have students finished the design scheme, to combine with the society projects having positive significance aims at making the students understand the object of design and further explore the society value of the design.

孙建华
CIID 副理事长
厦门天诺国际设计顾问机构的创办人兼设计总监

"6+1"项目已开始了数年。两个月前，应学会之邀在广州第一次参加毕业设计答辩评审会，近距离了解了这一活动的最新情况；同时，作为一个侧面，间接了解了国内大学设计专业学生的所学、所思、他们的专业能力与综合素养。

一天的设计答辩时间颇为紧张。来自不同院校的同学明显在分析工具、设计关注、表达方式、思维价值观上有很大不同。针对同一课题，环艺（室内）、景观、展览3个专业的同学从不同的角度进行思考、展开，有分工又有小组协作。综合他们的成果，惊喜发现相对完整地对实践课题作出了圆满的设计解决方案。

这一活动的意义毫无疑问是积极的，我认为至少有下述几个方面值得肯定：

（1）由于校企联合的优势，选择的课题是实际项目。更为可贵的是，据了解，每年选题均面向年度设计比较关切的方向，这样，对学生了解校园之外的宏观市场、设计业界动向非常有帮助；

（2）跨校联合的活动，让参与活动的高校院系之间、学生之间、教师之间形成非常好的互相学习借鉴。而且由于活动时间跨度长、有系统性，这种交流不是碎片式的，能够真正起到对设计教育的促进发展与提高的作用。

（3）CIID设计学会的专业支持，可以充分运用其专业、高端的设计资源起到推动、指导、引领作用，让设计教育面向设计社会实操层面，让学生更近距离接触到未来的设计实践工作模式。

当然，在评审中也看到设计小组个体之间的专业技能与最终成果质量的差距，但我想客观反映这种差距也是学习成长的另一重要途径。

向"6+1"活动各位发起、组织、参与的老师同学们致敬，并祝学会这项活动越办越好！

The "6+1" project has been started for many years. Two months ago, I participated in the graduation design defense comments for the first time in Guangzhou upon the invitation and thereby had the chance to learn the latest situation of this activity at a close distance, and at the same time, as a side, directly know what the students that major in design profession of domestic universities study and think about, and their professional capability and comprehensive quality.

The time of whole day for design defense is very nervous. Students from different colleges are obviously different on analyzing tools, design attention, expression manner, and thinking values. For the same topic, the students from three professions including environment art (interior), landscape and exhibition think and expand from different angles; they divide the work and have small groups for cooperation. By synthesizing their results, I find with surprise that they relatively and completely make successful design solution for the practice topic.

The meaning of this activity is undoubtedly positive and I think the following several aspects must be affirmed:

(1) In view of advantages of college-enterprise cooperation, the topic to be chosen is the real project; and more precious, the topic of every year faces the direction that draw more attention of the annual design by understanding, as such, it is very helpful for the students to know macroscopic market and tendency of design field outside the college.

(2) The college-cross cooperation activity enables forming of very good studying and learning from one another between different departments, students and teachers of the colleges and universities. Since the time of activity has a long span and systematicness, such communication is not a fragment and can really act to promote the development of and improve the design education.

(3) By the profession support by CIID national society, we can sufficiently apply the professional and high-end design sources to act to promote, guide and lead, and make the design education face the real operation level of design society and make the students contact the future design practice work mode in a more close distance.

Of course, we find a difference between professional skills and the final result quality between individuals of the design group, but, the objective reflection of this kind of difference is also another important path for studying and growing. I pay my respects to teachers and students who start, organize and participate in the "6+1" activity and wish a good future for such a kind of activity held by the institute.

余平
CIID 常务理事
西安电子科技大学工业设计系副教授

由中国建筑学会室内设计分会主办的"室内设计6+1"校企联合毕业设计答辩活动，是一次高校与高校之间高水准的交锋，也是一次难得的大学生专业交流活动，更是一次高校与设计企业之间的对话。

答与辩的交锋是紧张而激烈的。答辩给学生提供了一个很好的锻炼机会与平台，相互学习、相互启发；对学校、企业、导师而言，也是一次实战演习，一次很好的提高机会；也给了我一次学习机会。

面对一个实地项目的命题，在限定条件下，各校从城市地域分析，历史建筑现场分析，到空间创意，商业定位，材料组织分析等拿出了各自的设计方案。每个学校都有自己的特点，有的注重对建筑的分析，有的注重对内部空间的分析，有的注重对业态的分析。6所学校各有发挥，也反映出一定的差距。

从总体上看，存在的主要问题是：对特定历史建筑认识不足，挖掘深度不够，如建筑的特征、门窗可开启的条件、建筑墙体上的特殊构件、历史遗痕等。造成创意流于平庸，似乎只要有一个空间就可以随心所欲地做任何创意。因此，在历史建筑的限定性条件下解决问题的能力有待提高，"限定"才能产生创意，也是激发创造力的源泉。

The "interior design 6+1" college-enterprise cooperation graduation design defense activity hosted by interior design branch of China construction institute is a confrontation of high level between college and college, and is also a precious professional communication activity of graduates, and is even a conversation between the college and the design enterprise.

The confrontation between answering and defense is nervous and drastic. The defense provides a good exercise chance and platform for the students so they can study from one another and get inspiration from one another. It is also an actual combat for college, enterprise and mentor and a good chance for improvement.

Facing the assigning of topic of an on-site project, the colleges present their own design schemes under the limited conditions from city terrain analysis, historical building on-site analysis to space creation and business position and material organization analysis. Each of the colleges has own features, some of them focus on analysis of building, some of them focus on analysis of interior space and some of them focus on analysis of commercial activities.

Six colleges display differently and this reflects a certain difference between them.

In general, the main problem existed therein include: insufficient acquaintance of the specific historical building, insufficient digging depth such as feature of the building, the condition depending on which the door and window can be opened, special members on the building wall body, historical trace and so on. Therefore, the creation tends to fall flat; it seems that only one space is enough for any creation at will. Thereby, the capability of solving the problem under the limitative condition of historical building should be improved; limitation can generate creation and is also a fountain that stimulates creative power.

庞伟
广州土人景观顾问有限公司首席设计师

整整一天的活动参加下来，有感慨、有遗憾、有兴奋、有敬意。同学们和他们的老师们是试图为当代和历史找出更多思考路径和现实出路的人，他们需要在工作中整合太多作为个体无论从跨度到厚度都不易把握的内容。我们看到了设计作为一种角度、一种工作、一种策略和一种力量，是如何成立和绽放的，是如何质疑和建树的，是如何在设计者和他们面对的城市对象之间建立起一种可信任的展望和期待的。

这是寄语，并为本次和下次活动鼓掌。

By participating in the whole day's activity, I have many special feeling; have regrets, excitement and respects. Students and their teachers try to find out more thinking paths and real outlets for the modern times and history. They need to integrate much more contents that the modern people cannot easily master no matter from span or thickness in the work. We recognize how the design is established and bloomed, is doubted and contributed, as an angle, a kind of work, a kind of strategy and a kind of strength, and how the design establishes a kind of reliable prospect and expectation between the designers and the city object they faced.

This is my note and let's applause for this activity and for the activity of next time.

冼剑雄
广州瀚华建筑设计有限公司建筑专业董事、总建筑师

很久没有回母校了，感谢主办方的邀请，使我有机会在红楼参加第三届CIID"室内设计6+1"校企联合毕业设计评审活动。虽然平日也经常参加很多的评审活动，但这一次的感受确实有所不同。来自全国的7所院校对同一个实际案例从不同的角度、用不同的方式来表述自己的构思，让人很有新鲜感。应该说这样的活动很有意义，也办得很成功。在这次活动中，既体现出了各院校不同的教学风格，也充分展示了同学们各异的思维角度和表达特点。让评委老师和同学们在活动中能充分交流和互相学习，这本身就是一种非常好的教育方式，值得坚持和推广。

这次参评的很多方案不仅能从设计的角度去构思，还能结合自己在日常生活中的体验来考虑实际功能的需求，十分难能可贵。其中有些同学的答辩表述非常精彩，让我印象深刻。尽管从这些设计方案中也能发现存在的一些欠缺，但通过这样的方式让院校之间横向交流学习，互相取长补短，对提高中国设计教育水准有很好的促进作用。

祝贺这次活动圆满成功，也祝愿活动继续越办越精彩！

There is a Long time that I haven't back to my old college. Thank you for invitation from the sponsor, so I have this chance to participate in 3rd CIID "interior design 6+1" college-enterprise cooperation graduation design review activity. Although I often participate in many review activities, the feel of this time is indeed different. Seven colleges from the country express own concept for the same one actual case from different angles with different manners, it gives a feeling of freshness. Such activity is very meaningful and is held successfully. The activity not only embodies different teaching styles of different colleges but also shows different thinking angles and expression features of students. The judge teachers and students can communicate with each other and study form one another during the activity, which is very good kind of teaching manner and worthy of being insisted and popularized.

It is very precious that many schemes for the review are designed from the angle of designing and take requirements of actual function into consideration by combining with one's own experience in daily life. The defense of some students is very wonderful and gives me deep impression. Although there are still some defects in the design schemes, through this manner, the colleges can communicate with each other to study from each other, which serve to improve Chinese design education standard.

I wish a successful completion of this activity and hope to hold more and more wonderful activities.

花 絮 Tibits

命题研讨过程

调研过程讲解

师生现场踏勘　　　　　　　　方案交流现场

师生热烈研讨

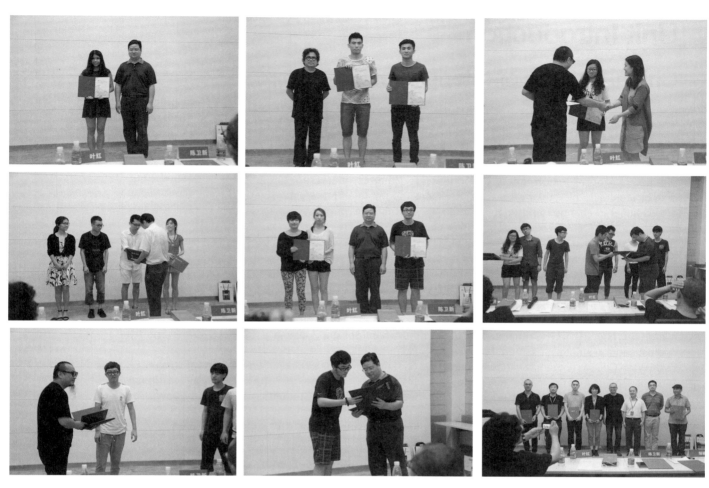

表彰奖励仪式

专家导师合影

Unit Introduction

China Institute of Interior Design

Six colleges display differently and this reflects a certain difference between them. The predecessor of China Institute of Interior Design (CIID) is China Institute of Interior Architects. Since itsestablishment in 1989, CIID has been the only authorized academic institution in the field of interior designin China.

CIID aims to unite interior architects of the whole country, raise the theoretical and practical level of China'sinterior design industry, pioneer the Chinese characteristics of interior design, help interior architects play their social role, preserve the rights and interests of interior architects, foster professional exchanges and cooperation with international peers, so as to serve and facilitate the construction of China's modernization.

Since its foundation 20 years ago, CIID hold abundant and colorful academic exchanges every year, building aplatform for designers to communicate and to study meanwhile update designer information of design industry, relatedcompetitions and business promotion, to enhance the better and rapid development of interior design industry ofChina.

Members of CIID are composed of individual members (including student members, associate members, fullmembers, advanced members, foreign members) and group members. By now, CIID has a large membership of morethan ten thousands prominent designers who are from all over the country and passed the strict assessment by CIID.

Every year CIID will organize various types of competitions which include Institute Award of China Interior DesignAward, Influential People of China Interior Design, "Renewal Design" Original Competition, National-level InteriorDesign Competition for Young Students, China Hand-drawn Art Design Competition and so on.

CIID Secretarial is located in Beijing, taking the charge of institute work. CIID secretariat publishes membershipperiodical china interior, Collection of Entries of China Interior Design, periodical Ornament and Decoration World,Home Adornments, ID+C. CIID website: www.ciid.com.cn

Tongji University

Tongji University, established in 1907, is a top university of China Ministry of Education. During the time ofrestructuring of the university and college systems in 1952,the Department of Architecture was formed at TongjiUniversity , and in 1986 was renamed as the College of Architecture Urban Planning (CAUP). Currently CAUPhas three departments: the Department of Architecture, the Department of Urban Planning, the Department ofLandscape Design. The undergraduate program covers: Architecture, Urban Planning, Landscape Design, HistoricBuilding Protection and Interior Design. CAUP is one of China's most influential educational institutions with the mostextensive programs among its peers, and the largest body of postgraduate students in the world. Today, CAUP hasbeen recognized as an international academic center with a global influence in the academic fields.

Tongji University's interior design education originated from the Department of Architecture which started toconduct interior space research in the 1950's.In 1959, it applied for the establishment of the "Interior Decoration andFurniture Specialty" within Architecture Discipline. In 1986,approved by the Ministry of Education and the Ministryof Construction, the "Interior Design Discipline" was formally founded. Starting to admit undergraduate studentsin 1987,Tongji University was one of two earliest high education institutions in mainland China to train interior design professionals in a University of science and technology. In 2011,

"Interior Design" officially became the secondary discipline of the Architecture Discipline. In the same year, the "Interior Design Research Team" was established,providing even broader room for subject development. Tongji University's interior design education crystallizes its own characteristics, emphasizing rational thinking andproposing the interior design concept of "human centric, ecological consciousness, overall environmental perspective,equal time and regional characteristic significance, technology and art integration"

单位介绍

CIID 中国建筑学会室内设计分会

中国建筑学会室内设计分会（简称CIID），前身是中国室内建筑师学会，成立于1989年，是在住房和城乡建设部中国建筑学会直接领导下、民政部注册登记的社团组织。CIID是获得国际室内设计组织认可的中国室内设计师的学术团体，是中国最具权威的室内设计学术组织。

学会的宗旨是团结全国室内设计师，提高中国室内设计的理论与实践水平，探索具有中国特色的室内设计道路，发挥室内设计师的社会作用，维护室内设计师的权益，发展与世界各国同行间的合作，为我国现代化建设服务。

CIID成立20多年来，每年举办丰富多彩的学术交流活动，为设计师提供交流和学习的场所，同时也为设计师提供丰富的设计信息，提供各类大型赛事信息，提供各项商务帮助，促进中国室内设计行业更好更快地发展。

CIID设有个人会员（包括学生会员、准会员、正式会员、资深会员、外籍会员）和团体会员。目前，已有会员1万余名，均是经过严格资格评审的精英设计师，遍布全国各地。CIID每年举办各类赛事，包括中国室内设计大奖赛"学会奖"，"中国室内设计影响力人物"评选，"设计再造"创意大赛，"新人杯"全国青年学生室内设计竞赛，中国手绘艺术设计大赛等一系列奖项。

CIID秘书处设在北京，负责学会相关工作。秘书处定期出版会员会刊《中国室内》以及《中国室内设计年刊》，同时学会拥有会刊《装饰装修天地》《家饰》《室内设计与装修ID+C》。CIID的官方网站为中国室内设计网。

同济大学

同济大学创建于1907年，教育部直属重点大学。同济大学1952年在国家院系调整过程中成立建筑系，1986年发展为建筑与城市规划学院，下设建筑系、城市规划系和景观学系，专业设置涵盖城市规划、建筑设计、景观设计、历史建筑保护、室内设计等广泛领域。同济大学建筑与城市规划学院是中国大陆同类院校中专业设置齐全、本科生招生规模最大，世界上同类院校中研究生培养规模第一，具有全球性影响力的建筑规划设计教学和科研机构，是重要的国际学术中心之一。

同济大学室内设计教育起源于建筑系，同济大学建筑系于20世纪50年代就开始注重建筑内部空间的研究，1959年曾尝试在建筑学专业中申请设立"室内装饰与家具专门化"。1986年经国家建设部和教育部批准，同济大学建筑系成立了室内设计专业，1987年正式招生，成为中国大陆最早在工科类（综合类）高等院校中设立的室内设计专业。1996年原上海建材学院室内设计与装饰专业并入同济大学建筑系；2000年原上海铁道大学装饰艺术专业并入同济大学建筑系。2009年同济大学开始恢复建筑学专业（室内设计方向）的招生工作。2011年建筑学一级学科目录下，设立"室内设计"二级学科。

同济大学建筑城规学院的教学理念为以现代建筑的理性精神为灵魂，以自主创造、博采众长的学术品格为本色，以当代技术与地域文化的并重交融为导向，以国际学科前沿的跟踪交流为背景。室内设计教学突出建筑类院校室内设计教学特色，强调理性精神，提出"以人为本、关注生态、注重环境整体观、时代性和地域性并重、融科学性和艺术性于一体"的室内设计观。

South China Uniersity of Technology

South China University of Technology is a key university that is directly affiliated to the National Ministry of Education. Located in Guangzhou, covered an area of three million square meters, SCUT is a comprehensive research university.Known for its engineering courses, SCUT also provides courses of science, laws, liberal arts, andbusiness to create a distinct academic environment for undergraduates and past graduates.

The School of Architecture has a long history. It grew out of Xiangqin University in 1932 and Faculty of Engineeringof National Sun Yat-sen University in 1938. It includes State Key Laboratory of Subtropical Building Science,Architecture Department, Urban Planning Department and Institute of Architectural Design Research. The School has built strong teams of education, research and construction consulting and academic research groups leaded byAcademicians He Jingtang and Wu Shuoxian.

Architecture is the most popular major in SCUT. It attracts lots of excellent students all over China and across theglobe every year. There are more and more undergraduate and graduate students and alumni that gain top prize ininternational competitions. Recently, about forty students each year gets to study abroad in top Architecture Schools, such as Cornell, MIT, Harvard, Columbia, U Penn, SCI-ARC, USC, Cambridge, UCL, TUdelft, ETH Zurich etc. Theschool has dual degree programs with universities in Italy, Australia and other countries.

Harbin Institute of Technology

Harbin Institute of Technology affiliates to the Ministry of Industry and Information Technology, and is among the first group of the national key universities to enter the national "211Project", "985 Project" and to start thecollaborative innovation "2011 plan". In order to train engineers, the Mid east railway authority founded the Harbin Sino Russian school in 1920, the predecessor of Harbin Institute of technology, which becomes the cradle of China's modern industry and technical personnel. The School has evolved into a distinctive, powerful, first class national key university, which is multidisciplinary, open, researchful and with international influence.

The discipline of Architecture in Harbin Institute of technology is one of the earliest architectural subjects in China, with more than 90 years' ups and downs. The school of Architecture has 4 undergraduate disciplines, including Architecture, Urban Planning, Landscape Architecture, Environmental Design, and 3 first-level disciplines, including Architecture, Urban and Rural Planning, Landscape Architecture, and secondary master's disciplines in Design. We have the first-level doctorate and master's authorization in Architecture, Urban and Rural Planning and Landscape Architecture, and secondary-discipline master's authorization in Design, and Post-doctoral Research Institute on architectural first-level discipline.With the cultural spirits of rigor and diligence,The school of Architecture has created a devoted, distinctive, qualified and dedicated teachers' team.We have gained distinctive and outstanding achievements in undergraduate teaching, postgraduate education and scientific research, and have formed our own academic characteristics in the Design of Public Buildings in Cold Region, Regional Architecture, Building Technology in Cold Region, Architectural History and Theory, Urban Planning and Designing in Cold Region and Environmental Design in Cold Region.

Xi'an University of Architecture and Technology

Located in the historical and cultural city Xi'an, covering an area of 4300 acres, Xi'an University of Architectureand Technology has beautiful campus environment and academic atmosphere. This university has quite a longhistory, which can be dated back to the Northern University, founded in 1895. Since then, in the higher education history of modern China, this university has been accumulating the first batch of disciplines essence in civil engineer-ing, construction and environmental class. In 1956, this university was named as Xi'an Institute of Architectural Engineering. In 1959 and 1963, it was renamed as Xi'an Institute of Metallurgy and Xi'an Institute of Metallurgy and Construction. On March 8, 1994, approved by the State Board of Education, it was renamed as Xi'an University of

华南理工大学

华南理工大学是直属教育部的全国重点大学，坐落在南方名城广州，占地面积约 300 万平方米，以工见长，理、管、经、文、法等多学科协调发展的综合性研究型大学。

华南理工大学建筑学院历史悠久，前身为创建于 1932 年的襄勤大学以及 1938 年并入的国立中山大学工学院。建筑学院由亚热带建筑科学国家重点实验室、建筑系、城市规划系、建筑设计研究院组成，现有一支实力强大的教学、科研和工程设计咨询队伍，已形成以何镜堂院士、吴硕贤院士等为学术带头人的科研团队。

建筑学是华南理工大学的龙头专业，连续多年招生排名全校第一，每年全国各地均有优秀的学生加入到这个学科中来。在校学生和毕业生获得国际竞赛大奖的越来越多。近年来，平均每年有 40 名建筑学院本科学生到国外深造，留学于康奈尔、麻省理工、哈佛、哥伦比亚、宾大、SCI-ARC、USC、剑桥、UCL、TUdelft、ETH 等名校，得到国外名校的充分认可。学院与意大利、澳大利亚等国大学建立双学位互认课程项目。

哈尔滨工业大学

哈尔滨工业大学隶属于国家工业和信息化部，是首批进入国家"211 工程""985 工程"和首批启动协同创新"2011 计划"建设的国家重点大学。1920 年，中东铁路管理局为培养工程技术人员创办了哈尔滨中俄工业学校——即哈尔滨工业大学的前身，学校成为中国近代培养工业技术人才的摇篮。学校已经发展成为一所特色鲜明、实力雄厚、居于国内一流水平、在国际上有较大影响的多学科、开放式、研究型的国家重点大学。

哈尔滨工业大学建筑学学科是我国最早建立的建筑学科之一，历经 90 余载风雨砥砺。建筑学院建筑学学科现有建筑学、城市规划、景观学、环境设计 4 个本科专业和建筑学、城乡规划学、风景园林学 3 个一级学科和设计学二级学科硕士点。已获得建筑学、城乡规划学和风景园林学一级学科博士、硕士授予权，以及设计学二级学科硕士授予权，还设有建筑学一级学科博士后科研流动站。建筑学院始终秉持严谨治学、精于耕耘的文化精神，打造了一支朴实敬业、有特色、有能力、肯奉献的优秀教师团队。在本科教学、研究生培养及科学研究方面，特色鲜明，成绩显著。在寒地公共建筑设计、地域建筑设计、寒地建筑技术、建筑历史与理论、寒地城市规划与城市设计、寒地环境艺术设计等诸多方向上，均形成自己的学术特色。

西安建筑科技大学

西安建筑科技大学坐落在历史文化名城西安，学校总占地 4300 余亩，校园环境优美，办学氛围浓郁。学校办学历史源远流长，其办学历史最早可追溯到始建于 1895 年的北洋大学，积淀了我国近代高等教育史上最早的一批土木、建筑、环境类学科精华。1956 年，时名西安建筑工程学院。1959 年和 1963 年，曾先后易名为西安冶金学院、西安冶金建筑学院。1994 年 3 月 8 日，经国家教委批准，更名为西安建筑科技大学，是公认的中国最具影响力的土木建筑类院校之一及原冶金部重点大学。

Architecture and Technology and was recognized as one of China's most influential civil engineering colleges and the key university of the former Ministry of colleges and the key university of the former Ministry of Metallurgical. Featured by civil engineering, construction, environment and materials science, engineering disciplines as the main body, Xi'an University of Architecture and Technology is a multidisciplinary university also with liberal arts, science, economics, management, arts, law andother disciplines. The university has 16 departments, 60 undergraduate programs so it can launch the first batch ofundergraduate enrollment. It also has the right to recruit students by recommendation and the right of implementationof Accelerated Degree. Undergraduate art and design program is the featured major in Shaanxi Province. Founded in April, 2004, Xi'an University of Architecture and Technology was established by the undergraduatesfrom the major of art design and photography and from mechanical and electrical engineering industrial design and the relevant teachers from newly established sculpture and other specialties. The current undergraduate majors in this college include art and design, industrial design, photography, sculpture, exhibition art and technology, with more than 1,200 undergraduate students. Art Design was named "national characteristic specialty", "provincial famous professional". This university has gathered many multidisciplinary researchers, including architecture, planning, landscape, etc. All these research teams have a long history of working towards the research of western region cultures, through undertaking many national and provincial funds subjects. The Arts College has actively organized (or as the contractor) the national academic, discipline-building meetings; inviting international and domestic famous professors to come for academic exchanges. It also has developed management approach, and set up a special fund to encourage young teachers and outstanding doctoral students to carry out academic exchanges and international (inside) collaborative researches. In the meantime it has established friendly and cooperative relations with the universities in Europe, Asia and other countries. The university has taken the overall development of students as its training objectives, the improvement of theoverall quality of them as the aim to focus on. Relying on various student organizations carrier and platforms, theuniversity has carried out various forms of extracurricular activities. And also it has focused on strengthening academicexchanges and interaction, inviting scholars, experts and celebrities to come to listen to the lectures and presentations,which can broaden the students' horizons, improve their knowledge structure and culture their spirits of science, technology and humanities. In other ways, the university organized the students to actively participate in academiccompetitions, and guided or encouraged students to engage in research activities, and many students have published various papers in the national magazines. The college has transferred departments, libraries, laboratories and paid multi-interactive efforts or work together to build a teaching-research-student trinity open experiment (work) platform. The graduates trained by the college have been welcomed by employers and the graduates are in short supply.

Beijing University of Civil Engineering and Architecture

Beijing University of Civil Engineering and Architecture (formerly Beijing institute of civil engineering and construction) was established in 1936 and is the only construction university in Beijing municipal colleges and universities.In1936,Beijing institute of civil engineering and construction was founded. In1977,Beijing institute of civil engineering and construction up graded to be undergraduate colleges and universities.In1982,Beijing institute of civil engineering and construction became one of the first B.A. rewarder's Authorities.In1986,Beijing instituteof civil engineering and construction become the authority to confer master's degreering.In2011,Beijing institute of civil engineering and construction was listed on the "educational training project of outstanding engineers" by the Ministry of Eduion. In2012,Beijing institute of civil engineering and construction was approved to serve national specific requirements "The project on architectural heritage protection theory and cultivation of doctoral degree for technical talents". In April 2013,Beijing institute of civil engineering and construction was renamed as Beijing University of Civil Engineering and Architecture(BUCEA) by the Ministry of Education.
BUCEA actively adapts to the development of higher education and the needs of urban and rural construction in the capital, adheres to the principle of "based in Beijing, facing the whole county, relying on construction industry, serving the urbanization construction ",and now has become a comprehensive university with distinctivecharacteristics which predominance of the disciplines of engineering science and other subjects including management, science, law, art etc. support each other and coordinate development, and is an important force in Beijing and even the national urban and rural construction.

西安建筑科技大学是以土木、建筑、环境、材料学科为特色,工程学科为主体,兼有文、理、经、管、艺、法等学科的多科性大学。学校现有16个院(系),其60个本科专业面向全国第一批招生,有权招收保送生,实行本硕连读。艺术设计本科专业为陕西省特色专业。

西安建筑科技大学艺术学院成立于2002年4月,是由建筑学院的艺术设计专业和摄影专业本科生、机电工程学院工业设计专业本科生和新成立的雕塑专业及各专业关教师组建而成。学院现有艺术设计、工业设计、摄影、雕塑、会展艺术与技术5个本科专业,在校本科生1200余人。艺术设计专业被评为"国家级特色专业""省级名牌专业"。学院集聚了包括建筑、规划、景观等在内的多学科的研究人才,学科团队长期致力于西部地区地域文化研究,承担了多项国家、省部级基金课题。艺术学院积极主办(承办)国家级学术、学科建设会议;邀请国际、国内知名教授来我校进行学术交流;制定管理办法,并设立专项基金,鼓励青年教师和优秀博士生开展学术交流、国际(内)合作研究,与欧洲、亚洲地区的多所大学建立了友好合作关系。

学院以学生全面发展为培养目标,注重学生综合素质提高,依托各类学生组织载体和平台,开展形式多样的课外活动。注重加强学术交流与互动,邀请学者、专家和社知名人士来我院举办讲座和专题报告,开阔学生视野,改善学生知识结构,培养学生的科技、人文精神。组织学生积极参与学科竞赛,指导、鼓励学生从事科研活动,在国内刊物上发表各类论文。学院调动教研室、资料室、实验室,多方互动,通力合作,构建了教学、科研、学生三位一体的开放性实验(工作)平台。学院培养的学生深受用人单位欢迎,毕业生供不应求。

北京建筑大学

北京建筑大学(原北京建筑工程学院)肇始于1936年,是北京市属高校中唯一的建筑类高等学校。1982年成为国家首批学士学位授予单位,1986年成为硕士学位授予单位,2011年成为教育部"卓越工程师教育培养计划"实施高校,2012年获批服务国家特殊需求"建筑遗产保护理论与技术"博士人才培养项目,2013年4月获教育部批准更名为北京建筑大学。

本校积极适应首都高等教育发展和首都城乡建设的需要,坚持"立足北京,面向全国,依托建筑业,服务城市化",现已成为以工学为主,工学、管理学、理学、法学、艺术学等学科相互支撑、协调发展的特色鲜明的多科性高校,是北京乃至国家城乡建设的重要力量。

本校设有11个学院和3个基础教学单位。现有博士学位授权一级学科点——建筑学;有硕士学位授权一级学科点12个、硕士学位授权交叉学科点1个、硕士专业学位授权类别/领域点9个;拥有一级学科北京市重点(建设)学科5个——建筑学、土木工程、测绘科学与技术、管理科学与工程、城乡规划学。现有34个全日制本科专业,其中教育部特色专业建设点3个——建筑学、土木工程、建筑环境与能源应用工程;加入教育部"卓越工程师教育培养计划"的专业4个——建筑学、土木工程、给排水科学与工程、能源与动力工程;北京市特色专业7个——建

There are 11 faculties and three basic teaching units in BUCEA. At present the school has 31 full-time under aduate including 3 Characteristic Specialties of the Ministry of Education, namely, Architecture, Civil Engineering ,Building Environment and Energy Engineering Applications, Building Environment and Energy Engineering Applications Water Supply and Drainage Science and Engine ering. Project Management Surveying and Mapping Engineering, Automation ,The school offers Doctoral Degree programs of Architecture for first-level disciplines, and 12 Masters Degree programs for first-level disciplines which could contain 55 second-level disciplines, and 1 interdisciplinary of Masters Degree programs, and 9 Master's degree authorization; There are 3 key disciplines by Beijing municipality for first-level disciplines namely Architecture Civil Engineering Surveying and Mapping Scien and Technology ,and 2 Key construction disciplines for first-level disciplines, namely, Management Science and Engineering ,Urban and Rural Planning .At present.BUCEA has 1020 faculty and staff, including 710 full-time instructors. There are 588 full-time ins with postgraduate degree ,accounting for 83%,281 full-time instructors with doctoral degrees accounting for 40%,and 410 full-time instructors with high professional title including 108 full proflis.

The school is divided into 2 campuses, located in xicheng and daxing. The library, constructed by BUCEA and Urban rural Ministry of Construction, is the most complete library of architectural books Over the year, BUCEA has provided over 50000 excellent graduates for the country, who have become the backbone of urban and rural construction and management system of the capital. They are involved in major urban construction engineering in Beijing over the past 60 years, and a larger number of excellent talents emerged, and graduate quality is widely recognized in society.

The school actively participated in 2011 play, namely, Play of Promoting the Innovation Ability for Colleges and Universities, and developed deep cooperation with scientific research institutions and enterprises to carry out important scientific research projects which contain 12 provincial key laboratory, engineering research center social science base and science and technology park project of unit including key laboratory of the Ministry of Education constructed by urban rainwater drainage system and water environment engineering research center of the Ministry of Education for reprint active buildings and ancient tar architecture database key laboratory of national geographic information bureau of surveying and mapping, Beijing's research and lent training base for climate change and so on. The school was identified as one of the first two pilot universities for reform of equity incentive in Zhongguancun national innovation demonstration zone. In the recent five years, the school won 56 awards of science and technology, including 8 awards of science and technology progss of the State and invention and won the Second Await the Scientism and Technological progress of the State with the first unit for 3 years (2010-2012). BUCEA adheres to open running and promotes broad international cooperation and exchange. At present BUCEA has established intercollegiate exchange and cooperation relationship with 35 universities from 23 nations and regions.

The school will deeply carry out the strategy of Standing with quality prospering with talents, developing with Science and technology, running with opening and Strive to build a distinctive university of architecture with high level for the future ten years

Nanjing University of the Arts

Nanjing University of the Arts is one of the earliest arts institutions in China. It consists of14 schools, 27 undergraduate majors and 50 major directions. It has Master's and Doctoraldegrees and Post-doctoral stations in 5 subdisciplines under the first-class discipline of thearts: Arts Theory, Music and Dance, Drama and Film, Fine Arts Theory and Design Theory.

The professional background: in 2005, Display Design was set up as a major direction in undergraduate level and a research direction in master program in Nanjing University of the Arts; in 2008, it was incorporated into industrial design major as its one direction in School of Industrial Design; in 2011, it was approved by the Ministry of Education as an independent sub-discipline in then national disciplinary classification in undergraduate education; in 2012, it was classified into the first-class discipline of design with a new major name of "Art and Technology " .Through nearly 10 years of efforts by adhering to the principle that is students centered, academy-oriented, practice focused and development-guided, Art and Technology (Display Design) major has formed a coherent and open modular curriculum system of coherent knowledge and rational structure supported by modernization
and globalizationoriented course contents. The major is to cultivate professional design talents for the cultural sector, the museum sector, medium and large exhibition halls, the design community, the tourism sector, exhibition and other institutions. The graduates of this major are to have the ability to do design and research in the manner of integrating question, market and culture. And they are also to be cultivated as talents with the capacity of high-level artistic formation and excellent expression, as well as rational expertise structure and outstanding professional features.

筑学、土木工程、建筑环境与能源应用工程、给排水科学与工程、工程管理、测绘工程、自动化。

本校分为西城和大兴两个校区。图书馆与住房和城乡建设部共建有中国建筑图书馆，成为全国建筑类图书种类最为齐全的高校。多年来，本校为国家培养了6万多名优秀毕业生，成为首都城乡建设与管理系统的骨干力量。他们参与了北京60年来重大城市建设工程，校友中涌现出了一大批优秀人才，毕业生质量得到社会的广泛认可。

本校积极参与2011计划——"高等学校创新能力提升计划"，同科研机构、企业开展深度合作，联合开展重大科研项目攻关。现有城市雨水系统与水环境省部共建教育部重点实验室、代表性建筑与古建筑数据库教育部工程研究中心、现代城市测绘国家测绘地理信息局重点实验室、北京市应对气候变化研究及人才培养基地等20个省部级重点实验室、工程研究中心、社科基地和大学科技园项目。被确定为中关村国家自主创新示范区股权激励改革工作首批试点的两所高校之一。近5年来，本校获省部级以上科技成果奖励56项，其中荣获国家科学技术进步奖、科技发明奖共8项，2010、2011、2012连续三年以第一主持单位获得国家科技进步二等奖。2013年完成科技服务经费突破2.6亿元，连续7年过亿。

本校坚持开放办学，广泛开展国际教育交流与合作。目前已与24个国家和地区的38所大学建立了校际交流与合作关系。

南京艺术学院

南京艺术学院是我国独立建制创办最早并延续至今的高等艺术学府。下设14个二级学院，27个本科专业及50个专业方向。拥有艺术学学科门类下设的艺术学理论、音乐与舞蹈学、戏剧与影视学、美术学以及设计学全部5个一级学科的博士、硕士学位授予权及博士后科研流动站。

南京艺术学院从2005年开设了展示设计本科专业和硕士专业研究方向；2008年该专业并入工业设计学院，2011年会展艺术与技术专业作为独立的二级学科获得国家教育部的正式批准，2012年该专业又被归为设计学类，成为"艺术与科技"专业。南京艺术学院工业设计学院的艺术与科技（展示设计）专业以学生为中心，以学术为导向，以实践为手段，以发展为目标，通过近10年的发展，已经逐步形成知识融贯、结构合理、连贯而开放的模块化专业课程体系和走向现代化、全球化的课程内容。旨在为文化部门、博物馆部门、大中型展馆、设计团体、旅游部门、会展机构等单位培养具有一定的理论素养，专业知识合理，专业特点突出，具备问题导入、市场导入和文化导入的整合设计和研究能力，以及高度艺术造型及表达能力的专业设计人才。

Zhejiang University of technology

Zhejiang University of technology is a key comprehensive college of the Zhejiang Province; its predecessor can be traced back to the founding in 1910 as Zhejiang secondary industrial school. After several generations' hard working and unremitting efforts, the school now has grown to be a comprehensive University in teaching and researching which is very influential. The comprehensive strength ranks the top colleges and universities. In 2009, Zhejiang province people's government and the Ministry of education signed a joint agreement; Zhejiang University of Technology became the province ministry co construction universities.

In 2013 the Zhejiang University of Technology led the construction of Yangtze River Delta green pharmaceutical Collaborative Innovation Center which was selected for the national 2011 program, to become one of the first 14 of 2011 collaborative innovation center. There are 68 undergraduate schools; 101 grade-2 subjects of master's degree authorization; 25 grade-2 subjects of doctor's degree authorization; 4 postdoctoral research stations. Subjects include philosophy, economics, law, education, literature, science, engineering, agriculture, medicine, management, arts and other 11 categories. School teacher is strong. There are 2 Chinese academicians of Academy of Engineering, sharing 3 academicians of Chinese Academy of Sciences and Academy of Engineering; 6 national young experts with outstanding contributions; 3 National Teaching Masters, 3 winners of national outstanding youth fund, 2 people were selected to central thousand person plan, the Ministry of education,1 professor of the Yangtze River scholars, 1 innovative team of Ministry of Education, 2 national teaching teams, and 26 person were selected to all kinds of national personnel training plans. Zhejiang University of Technology adhere its motto "Profound accomplishment and invigorating practice. Accumulate virtues and good practice." To improve the quality of education in a prominent position, and strive to cultivate to lead, promote Zhejiang and even the country's economic and social development of elite talent.

Attention Historical Architecture and Culture Institute

Nanjing Guanzhu Historical Architectural Culture Research Institute is a social organization under the administration of Nanjing Press and Publication Bureau of Culture and Radio which engaged in non-profit social service activities. Our institute is committed to the research and arrangement of historical architectures and the culture of them. Thus, we put forward the judging basis of the "reasonable use" of historical architectures from cultural analysis.

浙江工业大学

　　浙江工业大学是一所教育部和浙江省共建的省属重点大学,其前身可以追溯到 1910 年创立的浙江中等工业学堂。经过几代工大人的艰苦创业和不懈奋斗,学校目前已发展成为国内有一定影响力的综合性的教学研究型大学,综合实力稳居全国高校百强行列。

　　2013 年浙江工业大学牵头建设的长三角绿色制药协同创新中心入选国家 2011 计划,成为全国首批 14 家拥有 "2011 协同创新中心" 之一的高校。目前学校有本科专业 68 个;硕士学位授权二级学科 101 个;博士学位授权二级学科 25 个;博士学位授权一级学科 5 个;博士后流动站 4 个。学科涵盖哲学、经济学、法学、教育学、文学、理学、工学、农学、医学、管理学、艺术学等 11 大门类。学校师资力量雄厚,拥有中国工程院院士 2 人、共享中国科学院和中国工程院院士 3 人、国家级有突出贡献中青年专家 6 人、国家级教学名师 3 人、国家杰出青年基金获得者 3 人、中央千人计划入选者 2 人、教育部长江学者特聘教授 1 人、教育部创新团队 1 个、国家级教学团队 2 个、各类国家级人才培养计划入选者 26 人次。浙江工业大学坚持厚德健行的校训,把提高教育质量放在突出位置,努力培养能够引领、推动浙江乃至全国经济和社会发展的精英人才。

南京观筑历史建筑文化研究院

　　南京观筑历史建筑文化研究院是南京市文广新局主管下的从事非营利性社会服务活动的社会组织。本院致力于对南京历史建筑的梳理,以及对各历史建筑文化的整理。由此,从文化分析的角度对历史建筑的 "合适性使用" 提出判断依据。

致谢
Acknowledgement

南京观筑历史建筑文化研究院
Attention Historical Architecture and Culture Institute
同济大学建筑与城市规划学院
College of Architecture and Urban Planning, Tongji University
华南理工大学建筑学院
College of Architecture, South China University of Technology
哈尔滨工业大学建筑学院
College of Architecture, Harbin Institute of Technology
西安建筑科技大学艺术学院
College of art, Xi'an University of Architecture and Technology
北京建筑大学建筑与城市规划学院
College of Architecture and Urban Planning, Beijing University of Engineering and Architecture
南京艺术学院工业设计学院
School of Industrial Design, Nanjing Unixersity of the Arts
浙江工业大学
Zhejiang University of technology
CIID 第十八（南京）专业委员会
The 18th （Nanjing）Professional Commettee of CIID
CIID 第五（西安）专业委员会
The 5th （Xian）Professional Commettee of CIID
CIID 第九（广州）专业委员会
The 9th （Guangzhou）Professional Commettee of CIID
CIID 副理事长 江苏省建筑装饰设计研究院院长兼总工 李宁女士
Jiangsu Provincial Architectural Design & Research Institute Co., Ltd. Li Ning
CIID 资深顾问 李书才先生
CIID Senior Consultant Li Shucai
CIID 副理事长 集美组设计机构创意总监 林学明先生
Guangzhou NEWDAYS creative director Lin Xueming
CIID 副理事长 厦门天诺国际设计顾问机构的创办人兼设计总监 孙建华先生
XiaMen ATENO International design consultancy Sun Jianhua
CIID 常务理事西安电子科技大学工业设计系副教授余平先生
Xidian University Associate professor of industrial design Yu Ping
广州土人景观顾问有限公司首席设计师 庞伟先生
Guangzhou Turen Landscape Planning CO.,LTD Chief Designer Pang Wei
广州瀚华建筑设计有限公司 建筑专业董事、总建筑师 冼剑雄先生
Guangzhou HANHUA Architects engineers CO.,LTD. Xian Jianxiong